C000233143

PERGAMON INTERNATIONAL LIB
of Science, Technology, Engineering and So
The 1000-volume original paperback library in aid
industrial training and the enjoyment of leisure
Publisher: Robert Maxwell, M.C.

THE MOTORWAY AGE

Road and Traffic Policies in Post-war Britain

THE MOTORWAY AGE

Road and Traffic Policies in Post-war Britain

by

DAVID STARKIE

PERGAMON PRESS

OXFORD · NEW YORK · TORONTO · SYDNEY · PARIS · FRANKFURT

U.K.	Pergamon Press Ltd., Headington Hill Hall, Oxford OX3 0BW, England
U.S.A.	Pergamon Press Inc., Maxwell House, Fairview Park, Elmsford, New York 10523, U.S.A.
CANADA	Pergamon Press Canada Ltd., Suite 104, 150 Consumers Rd., Willowdale, Ontario M2J 1P9, Canada
AUSTRALIA	Pergamon Press (Aust.) Pty. Ltd., P.O. Box 544, Potts Point, N.S.W. 2011, Australia
FRANCE	Pergamon Press SARL, 24 rue des Ecoles, 75240 Paris, Cedex 05, France
FEDERAL REPUBLIC OF GERMANY	Pergamon Press GmbH, 6242 Kronberg-Taunus, Hammerweg 6, Federal Republic of Germany

First edition 1982

Library of Congress Cataloging in Publication Data

Starkie, D. N. M. (David Nicholas Martin), 1942 –
The motorway age.

(Urban and regional planning series; v. 28)
(Pergamon international library of science, technology, engineering, and social studies)
Includes bibliographical references and index.
1. Roads—Government policy—Great Britain.
2. Traffic engineering—Government policy—Great Britain
I. Title. II. Series. III. Series: Pergamon international library of science, technology, engineering, and social studies.
HE363.G72S7 1982 388'.068 82-1641

British Library Cataloguing in Publication Data

Starkie, David
The motorway age.

(Urban and regional planning series; 28)
(Pergamon international library)
1. Roads—Great Britain.
2. Great Britain—Politics and government—1945 –
I. Title. II. Series.
388.1 HE363.G7

ISBN 0-08-026091 8 (Hardcover)
0-08-027924 4 (Flexicover)

Printed in Great Britain by A. Wheaton & Co. Ltd., Exeter

I'm now arrived — thanks be to the gods!
Thro pathways rough and muddy,
A certain sign that making roads
Is no this people's study;
Altho' I'm no wi' Scripture cram'd,
I'm sure the Bible says
That heedless sinners shall be damn'd
Unless they mend their ways.

ROBERT BURNS

Preface

AT present-day prices, a figure not far short of £20 billion has been invested in new or improved roads since the war: a huge programme of public works by any standards. But as a programme it was, and still is, shrouded in controversy. To some it has been "one of the great public investment projects of all time".[1] * To others it has been "one of the most costly mistakes in the nation's history".[2] And yet there has been no comprehensive account of the ebb and flow of ideas and pressures that gave rise to this controversial programme of road construction (or its recent decline). My object here is to fill that vacuum: to record the optimistic and enlightened feelings that accompanied the first motorway schemes, the subsequent transformation of the nation's mood, and the ensuing battles as road proponents confronted increasing opposition.

The book is written around a central theme which expounds the way in which road and traffic policies developed at the national level after 1945 in response to pressure to provide more "road space" both within and between towns. Consequently, the broad structure of the book is chronological – but not slavishly so; each chapter focuses on a particular issue or development so that there is an element of cutting back and forth in time from chapter to chapter.

The structure is as follows. The first chapter sets out the details of the initial motorway programme, which emphasized a need to speed traffic between towns. For the corresponding period (1945–1960), Chapter 2 covers developments, or more precisely the absence of developments, in towns and cities. The early concentration of resources on building new roads through the countryside had the result of aggravating urban problems, especially in the early 1960s. Chapters 3–7 record these problems and the response they called forth – the use of traffic management, the debate about road pricing and the long-term studies of traffic in towns. The next two chapters continue the urban theme, this time tracing events of the late sixties and early seventies when a marked reaction to the idea of urban motorways first developed. Chapter 10 presents the aftermath and continues the urban story through to the beginning of the 1980s. Chapter 11 is a transitional chapter returning the reader to the trunk-road theme by outlining the controversy on heavy lorries. The latter had a major

*Superscript numbers refer to References at end of text.

influence on the trunk-road and motorway programme in the seventies, and Chapter 12 considers this specific influence. Chapter 13 continues with the inter-urban theme and records the recent decline of the trunk-road programme and motorway proposals in particular.

Thus the historical sequence of the book starts with roads between towns, switches to consider roads and traffic within towns and then reverts to trunk roads. No doubt my critics could suggest a better order for these thirteen chapters, but I am trying to make my history readable as well as accurate, and there can be no perfect sequence.

In the final chapter I allow myself a greater degree of speculation. Here I examine the basic features of those road and traffic policies reviewed in the previous thirteen chapters, considering, for example, the length of time a basic policy lasted, what a change of policy appeared to be related to, and so on. And, from these observations, I try to infer something about the conditions that have given rise to changes of policy.

Many of the issues covered in the book are of course rather technical in nature. Nevertheless, I have made a particular effort to avoid the use of both jargon and excessive technical detail, and it is my hope that the book will interest the general reader as much as the specialist. References will be found at the end of the text (there is a separate list for illustrations) but notes elaborating a point appear occasionally at the bottom of the relevant page. There is also an appendix containing a list of Ministers and Secretaries of State associated with the transport portfolio.

Many people and organizations have contributed to the writing of this book in various ways. I have acknowledged this elsewhere, but here I would like to record my special thanks to a few: to Dr. Derek Scrafton for comments on a draft of the entire manuscript, to Barbara Baerenfaenger and Debbie Oakley for expertly interpreting my thumb-nail sketches and to Bronwen Jones for editorial advice. Most of all I would like to thank Kerry Clift and my secretary Mrs. Joan Turner. Kerry bit-by-bit transcribed to disk a manuscript which was then altered on numerous (a million?) occasions. Joan met the challenge of a thousand deadlines with equanimity, and the equal challenge of my handwriting with great efficiency. Finally, I would like to record my thanks to Peggy Ducker, Editorial Director at Pergamon Press, Oxford, for all her assistance over the years. I believe this is to be the last of many manuscripts that she has engineered; I wish her well in retirement.

Acknowledgements

I am most grateful to the following persons for helping in various ways:

David Banister; David Bayliss; John Black; Phillip Blake; Ron Botham; A. J. Carruthers; Saskie Hallam; J. W. Hedley; R. C. Jenkins; Tommy Lovelock; Ian Mackinder; John Phillips; Bill Proctor; Gabriel Roth; Barbara Shane; Robert Wilson; Robin Wilson.

I would also like to thank the following for the supply of illustrative material:

Birmingham City Council; Brims & Co. Ltd.; Department of Transport; John Murray (Publishers) Ltd.; Keystone Press Agency Ltd.; Leeds City Council; Leyland Vehicles Ltd.; Punch Publications Ltd.; Thames Valley Newspapers Ltd.; Times Newspapers Ltd.

Contents

List of Illustrations

CHAPTER 1

The First Motorways

IT was a cold but bright morning when in December 1958 Britain's first motorway was inaugurated, opening (as *The Times* put it) a new era in road travel in Britain. The achievement was a very modest one, a mere eight miles of bypass (around Preston, Lancashire) which was later to become an integral part of the M6 motorway linking England with Scotland. The opening ceremony was conducted by the Prime Minister, who commented that the motorway was a sign that Britain was determined under the Conservatives to adopt the most advanced of ideas.[1] With an election not far ahead, Mr. Macmillan, of course, was not slow to convey the impression that motorways were the Conservatives' peculiar contribution to a modernized affluent Britain. But the truth of the matter was not quite so simple – nor indeed quite so plural.

The Tea Room Plan

A dozen years before Macmillan cut the tape, the Preston bypass had appeared in the first official plan for a national motorway network. That was in the spring of 1946. Hopes then ran high for rapid recovery from the debilitating effects of the war, and the ideals of planning were being enshrined by the Labour Government in a series of legislative measures laying the foundations of a comprehensive system of urban and regional planning.

The first of these measures was the 1945 Distribution of Industry Act, which aimed to correct the serious pre-war imbalance of employment opportunities between the different regions of Britain. The mechanisms of correction were on the one hand controls on industrial expansion in the more prosperous South-East England and the Midlands and, on the other hand, financial inducements for attracting employers to the less prosperous "development areas" of Merseyside, North-East England, West Cumberland, South Wales and Central Scotland. Within regions another problem had been diagnosed, namely that of severe congestion and overcrowding in the major industrial cities, including London. Here the solution proposed was to decentralize some of the population to self-contained industrial "new towns" developed by independent planning corporations funded directly by the Treasury.

With industry increasingly dependent on motor transport since the First World War, better roads were seen as an important counterpart of such plans for reshaping Britain's economic and social geography. Consequently, Labour's proposals extended to roads too, and in May 1946 a formal link was forged between the planning legislation and highways when Alfred Barnes, the Minister of Transport, announced a road plan that "will form the framework upon which the planning of town and country will be based".[2] Thus, with Barnes' highway proposals announced in May, the New Towns Bill passed in November and the first new town designated simultaneously, a national master plan was rapidly taking shape.

It was envisaged that the highway plan would take ten years to complete. However, within the decade there were to be three phases of activity. The following two years were to see an emphasis placed upon clearing the serious backlog of road maintenance that had developed during the war; putting in hand the reconstruction of the blitzed cities and tackling accident blackspots. The following three years were to see the emphasis switch to some major works "including a limited number of motor roads [motorways] for the development areas" and some road works required to relieve city traffic congestion.

The third, final and most pregnant phase covered years six to ten inclusive. This was to see emphasis placed upon the "comprehensive reconstruction of the principal national routes". These for the most part, it was thought, would be better constructed as motorways, and Alfred Barnes arranged for a diagram depicting them as such to be exhibited in the Members' tea room in the House of Commons. The tea room plan focused on an "hour glass"-shaped network of motorways of roughly 800 miles in total length. These would link the West Midlands with Hull, South Lancashire, Bristol and London. The base was formed by a motorway between London and South Wales whilst a northern counterpart linked South Lancashire across the Pennines with Hull. The only exceptions to this tightly formed motorway network were a link between Manchester, the Wirral and North Wales, and a continuation of the north-western motorway from Birmingham past Preston towards the Scottish border. Complementing the motorways were a series of up-graded general purpose highways radiating out from London. The longest of these was an improved Great North Road, first to Newcastle and then on to Edinburgh where it linked with similarly improved roads to Glasgow, Inverness and Aberdeen.

To aim to complete this plan within ten years was undoubtedly ambitious, but, at the time of its announcement, the Government seemed aware of what it entailed. As Barnes noted, the Government was prepared to see expenditure expand in the first year up to a level of £80 million, far in advance of any pre-war figure. However, within a very short time general economic circumstances, which had never been good, deteriorated rapidly with the exhausting of post-war credits given by the U.S.A. and Canada. There was an urgent need from 1947 onwards to increase exports substantially, and with skilled labour, coal, steel and other materials in short supply, capital investment was cut back. Rather than spread the misery widely, however, the Government chose, in effect, to cut drastically certain proposed programmes whilst preserving the substance of others. Thus whilst sixteen "new towns" were designated between 1946 and 1950, the roads element in the *de facto* national plan collapsed. A number of major projects such as the Severn Bridge and the Tyne Tunnel, on which preparatory work had started, were cancelled and, in total, the 1948 programme of large works was reduced by 80 per cent. Consequently, expenditure on new and improved roads, instead of outpacing that of the pre-war period, continued to languish at levels barely a quarter of those reached in the late thirties.

The road situation altered little with the change of Government in 1951. Economic difficulties continued for a time, accentuated by the commodity shortages of the Korean War. Nevertheless, 1952 was to see also a new emphasis in economic policy. The Conservatives won the 1951 election with a platform of "Setting the People Free", and controls inhibiting consumer expenditure were, as a consequence, slowly dismantled. The emphasis now was on expanding consumer demand with public investment taking a hind seat. Consequently, whilst purchase tax on motor cars was cut in 1953, expenditure on new roads continued at levels no higher, indeed slightly lower in real terms, than those achieved by Labour between 1948 and 1950.

But private motor vehicles run on public roads and, as Boyd-Carpenter was to note in his memoirs, by the mid fifties: "Public anger was rising against the inadequacies of our road system and the inaction of the Government in respect to it."[3] It was a situation of which the roads lobby were to take advantage. In 1948 the British Roads Federation, then largely representing road construction interests, had joined forces with the Institute of Highway Engineers and the Society of Motor Manufacturers and Traders (SMMT) to press the Government for more expenditure on roads. But in the fifties the lobby was to broaden and intensify.

The Conservative Government, however, was in a curious dilemma. Even in these halcyon days before road schemes became a major environmental issue, new roads still took a considerable time to plan and prepare. The statutory procedures involved were long and intricate, and the total process for a major project was considered, even then, to take at least four years. Unfortunately for the Government, from 1947 until it first announced a roads programme in December 1953, preparatory work on road schemes had been restricted in order to avoid unnecessary expenditure. In fact, work had been limited to preparing "schemes"* or "orders" which merely protected the likely route of a motorway or trunk road project:† a job done largely in accordance with the 1946 plan.

*Schemes were prepared under the provisions of the Special Roads Act 1949. The Act made possible the building of roads such as motorways for the exclusive use of certain types of traffic. The first motorway line established under the Act (in 1949) was a 28-mile length of what is now the Bristol–Birmingham motorway.

†The Trunk Roads Act 1936 transferred to the Ministry of Transport responsibility for 4500 miles of road throughout Great Britain which would constitute a national system of through routes. By 1977 this mileage had grown to 9534, although responsibilities were shared with the Secretaries of State for Scotland (after 1956) and for Wales (after 1965).

This had two consequences. First, it meant that regardless of the degree of intent on the part of the Government it was impossible to start a large-scale construction programme until the latter part of the fifties. The second, and this time a rather peculiar and ironic, consequence was that if the Government wished to proceed as quickly as it could in preparing schemes for construction, then it had little option but to adopt elements from the 1946 Labour plan.

Thus the Government was in a cleft stick. The delay in getting schemes started – construction work on the Preston bypass, for example, did not begin until June 1956 – merely aggravated pressures for expanding the roads programme. During 1956 alone there were nine Parliamentary debates on road conditions. Outside Parliament, the Roads Campaign Council was formed and launched a "Roads Crusade" to marshal popular opinion in favour of better

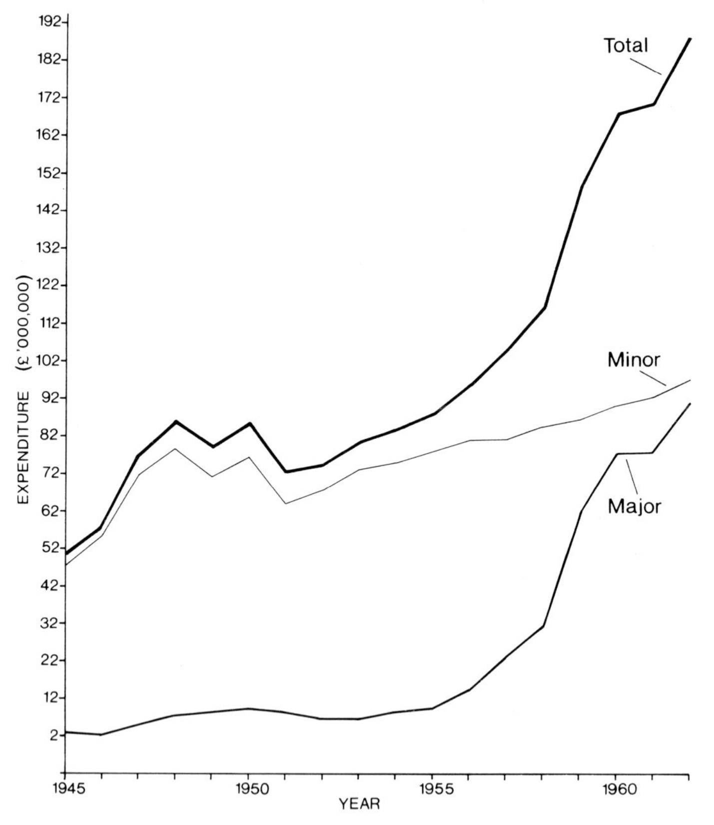

FIG. 1.1. Road expenditure 1945–62 (at constant 1962 prices): minor expenditures largely on maintenance increased immediately after the war, rectifying years of neglect and the damage sustained from bombs and military vehicles. Expenditure on new and improved roads on the other hand languished until the second half of the 1950s.

roads. The big corporate battalions also played their part in building up the pressures, with the TUC joining forces with employers' associations on the National Production Advisory Council to lobby its chairman, the Chancellor of the Exchequer, for more spending on roads.

Consequently, when the Government responded – as it did in July 1957 – it had to do so by digging deeper into its inheritance from 1946, namely the tea room plan. This time the indebtedness to it was to prove considerable. On this occasion the Minister of Transport, Harold Watkinson, announced an expanded programme containing a number of large projects that were all expected to start within four years. These projects were: the reconstruction of the Great North Road with five bypasses of motorway standard; a motorway from London to the North West (in effect a continuation of the London–Birmingham motorway already announced by that time); better roads to the Channel ports; an outlet from London to the West; and better communication between the Midlands and South Wales.

When placed on the map these schemes added up to a sizeable chunk of the 1946 plan.* During the ensuing debate Watkinson referred to this earlier plan as "a very good plan". He added that: "Today, of course, it is already out of date."[6] Fortunately for the Minister, no member in the House was inclined to draw attention to this latter remark and its implications for the Conservatives' own proposals.

The Drive for Top Gear

Having faced the consequences of there being too little in the pipeline on one occasion, the Ministry was determined not to let the same thing happen again. Consequently, before the first motorway opened, steps were being taken to advance the roads programme beyond Watkinson's quartet of major projects. In

*In fundamental terms the only deviation from Labour's plan was the inclusion of the Ross Spur motorway. This had its origins in a recommendation made in 1953 by a committee set up to study the communications of South Wales, and its early adoption in the programme probably represented an attempt to offer Wales something in view of a decision to press ahead with a new Forth road bridge in advance of a similar bridge over the river Severn. It had been the intention of the post-war Labour Government to give higher priority to the Severn Bridge which was seen as a "vital link in the Government's policy of developing the industrial areas of Wales and the port facilities of South Wales".[4] However, Scottish pressures in the mid fifties for a Forth Bridge were considerable. As Boyd-Carpenter recalled: "He [the Secretary of State for Scotland] demanded for what . . . were political reasons, that top priority in the new roads programme should be a road bridge over the Firth of Forth. The case on merits was poor, but with a General election not far away and the Conservative Party's hold in Scotland precarious, he could, and did, mount a formidable attack."[5] In 1957 the Ross Spur (M50) motorway was expected to cost around £6 million compared with £15 million for the cheapest Severn Bridge option. Therefore it was easier to give the former higher priority at an early date. In the late seventies the M50 was one of Britain's least used motorways. Average traffic flows were about a tenth of those on the busiest sections of the M1 and M4. The Forth Bridge opened in 1964, the Severn Bridge in 1966.

1957 a long-term planning group was established within the Ministry. The approach adopted by it was to encourage the road engineering divisions of the Ministry – there were nine of these for England and Wales, with Scotland forming another separate division answerable after 1956 to the Secretary of State for Scotland – to collect data on traffic flows, accident black spots and so on. The divisions then assessed, as a Ministry witness was to tell the Select Committee on Estimates in 1959, the influence of predicted future traffic on existing roads and, consequently, were able to build up a catalogue of schemes considered necessary to bring the trunk-road network up to a suitable standard for handling such flows. From these divisional assessments the Ministry was able to form a broad picture of priorities, in particular which main roads were overloaded and where the main industrial traffic was going.

Surprisingly, when the broad picture began to take shape it appeared to confirm to a remarkable degree that what was proposed already was basically correct and appropriate. The five major projects which since 1957 formed the backbone of the roads programme were "very strongly" endorsed by activities of the long-term planning group.[7] But, in addition to this confirmation of existing projects, at the turn of the decade the strategic planning exercise led to proposals for further extending the motorway network along the lines prescribed in the tea room plan.

The proposals were for motorways from London to Crawley new town, from London to Basingstoke (a designated overspill town for London's population), from London to Bishop's Stortford, the extension of the M1 towards Leeds, and for a central section of the M62. Of these schemes, the largest, the M1 extension and the M62, had appeared as motorways in the 1946 blueprint, whilst the London–Basingstoke and London–Crawley schemes had appeared also as improved trunk roads. Only the London–Bishop's Stortford proposal cut an entirely new line. It was a maverick scheme difficult to explain in terms of traffic congestion problems more severe than on other routes not graced by motorway proposals. But the proposed line did run close to a large wartime aerodrome called Stansted, soon to be designated as London's next major airport.

While the information for the long-term plan was being collected and collated in the late fifties and early sixties, other significant developments were taking place. The first batch of motorways were, of course, now opening. The Preston bypass in 1958 and the first 72 miles of the M1 in 1959 were followed in 1960 by the Lancaster and Maidstone bypasses (the latter part of the strategy of improving access to the Channel Ports) and the Ross Spur motorway. But in addition, substantial progress was being accomplished on other parts of the trunk road network. In 1959/60 and again in 1960/61 over 20 miles of improvements were completed on the Great North Road, so that by 1963 160 of the 280 miles between London and Newcastle consisted of dual carriageways. Elsewhere a number of high-quality bypasses opened.

These schemes had the effect of whetting the appetite of road users and thus strengthening the pro-road lobbies. But, in 1960, the case for building more motorways received a considerable boost with the publication of a now famous report calculating the economic costs and benefits of the London–Birmingham motorway.[8] The analysis was carried out by a joint team from the Road Research Laboratory and the University of Birmingham. It was done during the late fifties at a time when the motorway was still under construction. Therefore, it based its analyses on various assumptions concerning the likely speed of vehicles on the motorway, the degree of diversion from other routes to the faster motorway, and so on.

The research team first considered the savings in working time as a result of traffic transferring to the motorway, and the value that might be placed on these savings. Faster and smoother motoring also allowed for reductions in vehicle-operating costs and for fewer vehicles to do the same amount of work. These too were valued. From these benefits the team then offset the costs of additional vehicle-miles incurred by drivers diverting to take advantage of the motorway. They then estimated the benefits to traffic which would remain on the existing, but prospectively less congested, roads, and the benefits of fewer accidents. A final offsetting item was the maintenance cost of the new motorway. These various items gave three different estimates of the net annual savings from the motorway. Each estimate related to a different assumption about average speeds that could be expected on the motorway. To complete the final picture, these net savings were expressed as rates of return on the capital cost of the schemes and the rates were then inflated to allow for traffic growth and savings in people's non-working time.

The inclusion of a value for time saved by those travellers not using the motorway in the course of their work (but possibly going to and from work) required, of course, something of a stab in the dark. In fact three different values were considered ranging from 4 shillings to 8 shillings for an hour's saved time. This had the effect of considerably boosting the predicted rate of return of the motorway, which it was thought would then lie between 9·9 and 15·2 per cent in 1960, and between 17·6 and 27·3 per cent in 1965, well above the then prevailing interest rate of 5–6 per cent.

In spite of the tentative nature of the results of the London–Birmingham motorway cost benefit analysis, it appeared to have a considerable impact on thinking both in and outside Whitehall. For the roads lobby it was grist to the mill. For the Ministry it was a most convenient argument to put before Treasury at a time when it was seeking support for its revised long-term strategy.

The Treasury were persuaded by the Ministry's case. In July 1960 the first "rolling programme", whereby roads expenditure was to be fixed five years ahead and rolled forward each year, was announced. In addition, the financial amounts in this programme were increased significantly. The motorway programme was accelerated; Watkinson's five major projects were to be substantially finished by the mid 1960s and the long-term strategy was now to complete a total of 1000 miles of trunk motorway and 1700 miles of new or reconstructed dual-carriageway trunk road. At last the trunk-roads programme was shifting into top gear: a momentum it was to maintain for many years to come.

Summary and Conclusions

Britain's first plan for a national network of motorways appeared in 1946. At that time the plan was intended to complement the post-war Labour Government's other plans for changing the balance of employment and settlement within and between regions. But, as it transpired, economic circumstances soon resulted in choices having to be made between the Government's different programmes for capital investment. Consequently, it was the implementation of the roads plan that was postponed whilst resources were committed to "new towns" and to boosting manufacturing facilities in the development areas.

It was not until the Conservatives had been in power for a couple of years after 1951 that economic circumstances began to change for the better to an appeciable degree. When they did so, the rapid growth in traffic on the roads was paralleled by an equally rapid growth in pressures for an expanded programme of road investment. A striking feature of these pressures – particularly from the standpoint of recent years – was their broad base. The motoring lobbies were joined by employer and employee associations, and, within Parliament, Labour members joined Conservative backbenchers in putting pressure on the Government. There was a wide consensus, and the environmental and social issues that were to infuse the roads debate during the seventies received negligible attention.

It was in this heady atmosphere of consensus that a change of policy finally came. In a series of announcements between 1953 and 1955 the Government committed itself to an expanded roads programme. Construction, however, was a more hesitant matter. Little preparatory work had been carried out on major road schemes since the war, and the necessary procedures involved were complex and time consuming. With pressures on the Government continually building up, it was necessary to expand the programme around those schemes, basically those in the tea room plan, on which some preliminary work had been done already.

Whilst construction of these schemes was still in its early stages, preparatory work commenced on a much larger programme. During the late fifties a long-term planning group beavered away in the Ministry, collating data from various traffic surveys. By 1960 a picture began to emerge which confirmed as broadly correct those projects already started. More importantly it confirmed also as next in line, a number of additional schemes from the 1946 plan. However, largely as a result of upgrading a number of these to motorway standard, the tea room plan was now stretched to form a proposal for a basic 1000-mile network of motorways.

The release of these proposals in 1960 came at a time when the climate appeared ripe for yet another boost in roads spending. The London–Birmingham (M1) motorway, the first really long stretch of truly inter-urban motorway, had opened in a fanfare of publicity a few months before. Its driving qualities were quickly appreciated and, with its economic benefits estimated in an important cost benefit study to give a very good rate of return on capital spent, the Government and the Treasury appear to have been persuaded by the political and economic case for spending more on roads. Within a year of the opening of the M1, a "rolling" five-year expenditure programme had been outlined and completion of the 1000-mile motorway network – the modified tea room plan – had become the policy of the Tory Government.

CHAPTER 2

Town Roads: A Conservative Era

IN the late fifties, Britain had both a policy for building new roads and, as a concrete expression of this, almost 100 miles of motorway in use with nearly the same mileage under construction. But it was a construction policy focused exclusively on getting traffic more smoothly, efficiently and safely from one town to another. Within towns, it was a very different matter.

Here the Tories inherited from the post-war Labour administration a policy of discouraging urban motorway planning. As a consequence, the first plan incorporating a true urban motorway was not submitted for Government approval until 1960. Nearly five years later, by the time the Conservatives left power after nearly thirteen years of unbroken rule, Britain had around 300 miles of motorway. But the tapes had still to be cut on an inner-urban motorway designed specifically to solve a city traffic problem.

Post-war Blueprints

There were, of course, plans for some new roads during the fifties. But on the whole, the concept and scale of these were pretty anaemic by later standards. Most had been drafted as a consequence of the 1947 Town and Country Planning Act – described as one of the largest and most complex pieces of legislation ever passed by Parliament and, in fact, the cornerstone of the whole planning system created after the Second World War.[1] Of the many duties established by the Act, one of significance was the requirement placed upon all counties and county boroughs to prepare and submit a development plan for approval by the Minister responsible for planning. This plan was to be reviewed every five years and was to consist of a written statement and maps showing all important developments and intending changes in the use of land, including that used for highway purposes, over a future period of twenty years. The task was a mammoth one and it was to be well into the fifties before many local authorities were in a position of having an approved plan from which to work.

During the preparation of the initial plans in the late forties and early fifties, the highway component within them was to be influenced greatly by two particular factors. The first was what turned out to be extremely conservative official forecasts of the expected number of future vehicles. In 1945, for example, two vehicles forecasts were produced, one for the urban areas of the country and the other for rural areas. Both used actual 1933 vehicle registrations as a base, and it was anticipated that by 1965 vehicle numbers would have swelled by 75 per cent in towns and cities, and by a much lower 45 per cent in rural areas. As it transpired, by 1950 – five years later – vehicle numbers had nearly doubled compared with 1933. By 1954 a new set of forecasts had been issued, this time anticipating a 75 per cent increase in traffic over the forthcoming twenty years. But by 1962, only eight years later, this number had been reached already. In 1957 the Ministry of Transport and Civil Aviation forecast 8 million vehicles by 1960; by that time there were in fact 9 million.

The second major influence on the shape and form of the road proposals in

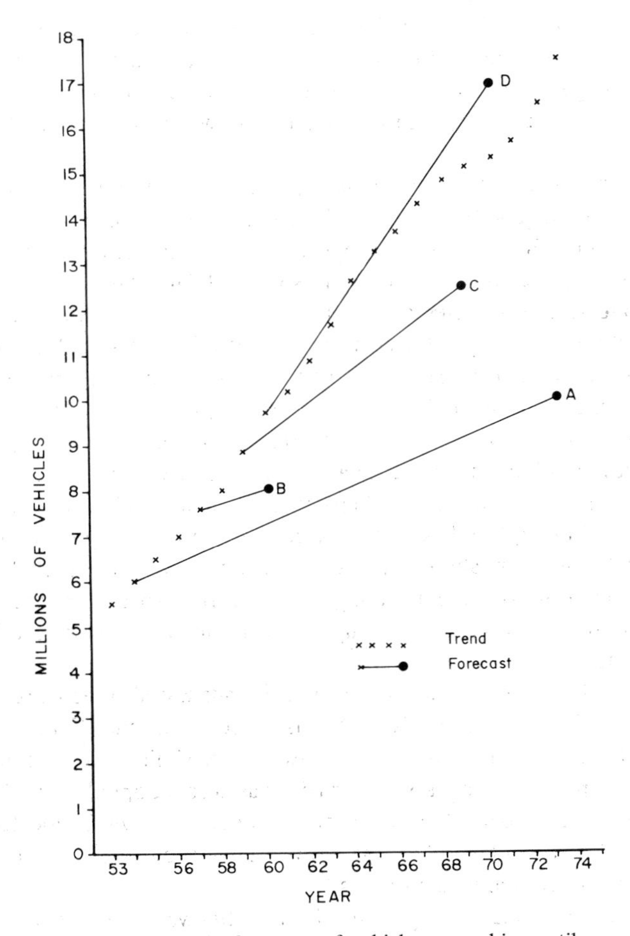

FIG. 2.1. Early forecasts of vehicle ownership: until 1960 official forecasts consistently and seriously underestimated future growth in the number of vehicles.

the development plans was an important report on the *Design and Layout of Roads in Built-up Areas,*[2] prepared at the end of the war. This report propagated the idea of a "ring and spoke" road pattern for cities. Translated in terms of an urban area of 250,000 people, this concept revolved around three ring roads cut across by a number of radials forming, in effect, spokes to a wheel. The primary function of the outer ring was to link the peripheral communities and to act as a distributor between radials. On this assumption, it was to be located as a rule not outside the town limits but within the outer fringe of present and potential development. Although designed for high speeds, it was to provide for all classes of traffic.

The intermediate ring on the other hand was to incorporate to a larger extent than the outer ring "existing roads with all the diversity of use to which the latter are subjected". Limitation of interconnection with subsidiary roads and of frontage access was to be aimed for but, except where entirely new sections were constructed, complete restriction was thought to be "out of the question".

The inner ring was a more interesting case. The central area within it was intended to be largely traffic free. Therefore, it was seen as important to exclude if possible the substantial proportion of central-area traffic made up of buses: "no adequate alleviation of congestion and danger is likely to be attained unless public service vehicles can . . . be diverted from the central area". However, the importance of good access to the centre by users of "mass transport" was recognized and, because 300 yards was considered a reasonable maximum walking distance, the inner ring, if at all possible, was not to exceed 600 yards in diameter. The ring was to become, in effect, a circular bus station, "becoming a more or less continuous picking up and setting down circuit" for bus passengers. Frontage access was to be restricted although not completely prohibited.

As for the linking "spokes", these were again seen as all-purpose roads with traffic signals and roundabouts at intersections, with possibly a grade-separated "fly-over" junction when connecting to outer ring roads. On the whole the "spokes" were seen to evolve through a "continuous . . . process of widening existing roads".

The possibility of subsurface and elevated roads was also recognized but only in highly developed parts of large towns or conurbations. Generally though, such structures were considered unnecessary. A similar view of motorways restricted to particular classes of traffic was also adopted. The Committee visualized that these roads "save in exceptional cases, . . . will be located outside, but reasonably adjacent to, the larger towns and cities, which will be connected to the motorway system by spur roads".

This then was the basic blueprint local authorities were to use, and to reinforce the point, the inner ring road "theory" as set out in the Committee's Report was incorporated by the Ministry of Town and Country Planning in its 1947 handbook *The Redevelopment of Central Areas.*

Such conservatism in Whitehall advice on urban road design, plus even more conservative traffic forecasts, appeared to kill most enthusiasm and enterprise amongst local authorities. During the fifties the development plans slowly emerged but few contained designs for complete inner ring roads. In many cases, the statutory plans (as approved by the Minister of Town and Country Planning) contained only bits of an inner ring programmed for completion sometime during a generous twenty-year period. Doncaster and Swansea, for example, included horseshoe-shaped road proposals but, at the other extreme, Reading proposed a mere 300-yard-long link road. There were exceptions of course to this piecemeal approach, but these were usually by those cities badly damaged by wartime bombing, like Coventry and Plymouth, or by the largest cities, like Birmingham, where activity was more concentrated and therefore traffic more intense.

In the latter case, civic pride, enterprise and advanced planning also played a major role. During the war, Birmingham had started to plan for the redevelopment of its commercial heart, and by July 1943 the City Council had approved in principle a city-centre scheme. This included proposals for an inner ring road which showed most of the characteristics later expounded as virtues in the Ministry's 1946 Report. The road was a divided carriageway, with provision in each direction for three lanes of moving traffic plus one lane of stationary vehicles. Together with large adjacent footpaths this gave the scheme a total width of 110 feet. Most of it was at ground level, although there was a short elevated section. To enable the purchase of the large areas of land required, legislation was promoted in the 1945/46 Parliamentary session. At the time, the road was described as "not quite a motor road in the terms of the experts, but a limited access road. That kind of road probably fits our traffic problems better."[3]

In a few cases where other authorities were willing to be more adventurous with their road proposals, the Ministry later disapproved. In London, for example, the London County Council Development Plan statement commented that "no single improvement of any great magnitude or of a novel kind such as a ring motorway around the inner area is included, because the Government in May 1950 made it clear that such a proposal would be unacceptable".[4] When the Plan was finally approved by the subsequent Conservative administration, it included a series of very modest schemes focused on improving the existing network of roads. The Elephant and Castle roundabout scheme, the widening of Park Lane, Notting Hill Gate and the Strand together with the extension of Cromwell Road were included in the list. But few such schemes were completed before the beginning of the sixties.

Elsewhere it was a similar story of very limited construction and conservative design. Even Birmingham's relative foresight paid few dividends. Though loan sanction for the acquisition of properties for their inner ring road was given by the Treasury in 1950, subsequent financial stringency and restriction on new

FIG. 2.2. Smallbrook Ringway, Birmingham: this section of Birmingham's inner ring road was constructed to wartime design principles – a three-lane divided carriageway with a fourth lane for parked vehicles, and adjacent footpaths and shop frontages. The photograph was taken in 1963. Later sections of the ring road (now called Queensway) were constructed to a modified design.

road construction meant that it was not possible to commence the road works until March 1957.

Getting By

For a time this lack of foresight and preparation was not of great consequence. Indeed, events as they unfolded possibly reinforced the Ministry's rather complacent view of the situation. Repairs to roads damaged during the blitz continued for many years after the war (expenditure on maintenance and minor improvements was nearly ten times that of new construction in 1953), and even the simple expedient of resurfacing turned out to have a surprising impact on street performance. Measurements during the early fifties in Glasgow, for example, had shown that resurfaced streets with fewer pot-holes could carry an increased flow of vehicles significantly faster.

But a large bonus came also with the continued demise and eventual removal of town tramway systems. Although the trams were replaced by buses, the latter were more manoeuvrable and less intimidating. The bus did not command its own track like the tram had done, and where buses replaced extensive tram networks the release of street capacity for other vehicles was quite considerable. For many large industrial cities throughout the fifties, this process of change from trams to buses provided a useful safety valve in the face of growing motor traffic. Birmingham's last tram ran in 1953, Edinburgh's in 1956, Liverpool's in 1957, and Sheffield's in 1960.

Thus in spite of the very rapid increase in traffic towards the end of the fifties decade (in three years urban traffic grew by about 40 per cent), by 1960 a surprising proportion of urban roads were still operating within their design capacities – capacities which, once reached, still gave a reasonable level of speed. Of so-called Class One roads, which formed the backbone of the urban street system, less than 50 per cent nationwide were "overloaded". Taking all main roads (trunk and Class One) in urban areas, the overloaded proportion was still only a fraction over a half. The picture varied a great deal, of course, from place to place and from road to road. But, in general, high levels of "over-

TABLE 2.1. *The Effect of the Removal of Trams on Traffic Speeds in London*[5]

	Roads with trams in 1950 but not 1952 (9 miles)			Road without trams in 1950 and 1952 (25 miles)		
	1950	1952	Change (%)	1950	1952	Change (%)
Average number of vehicles per hour	1130	1235	+9	1460	1480	+1
Average journey speed (m.p.h.)	12·7	14·2	+12	9·8	10·5	+7

FIG. 2.3. Road capacity shared with trams: this photograph shows early morning traffic crossing Blackfriars Bridge in the autumn of 1948. Private cars are conspicuously absent. A sizeable proportion of the road was dominated by the passage of tramcars.

load" on urban roads occurred only in the Midlands and in an area to the east of a line between the Solent and the Wash. At the other extreme, Scotland experienced a negligible problem whilst the North, Wales and South-West England fell in between.

Consequently, a crisis, if a crisis existed, was really confined at this time to the Capital and (in spite of the general Scottish picture) possibly to Glasgow, where the trams lingered on into the sixties. London's trams had disappeared by 1952 (they had not in any case intruded much into the central area) and the street space released was quickly absorbed. Here the pressures came from the

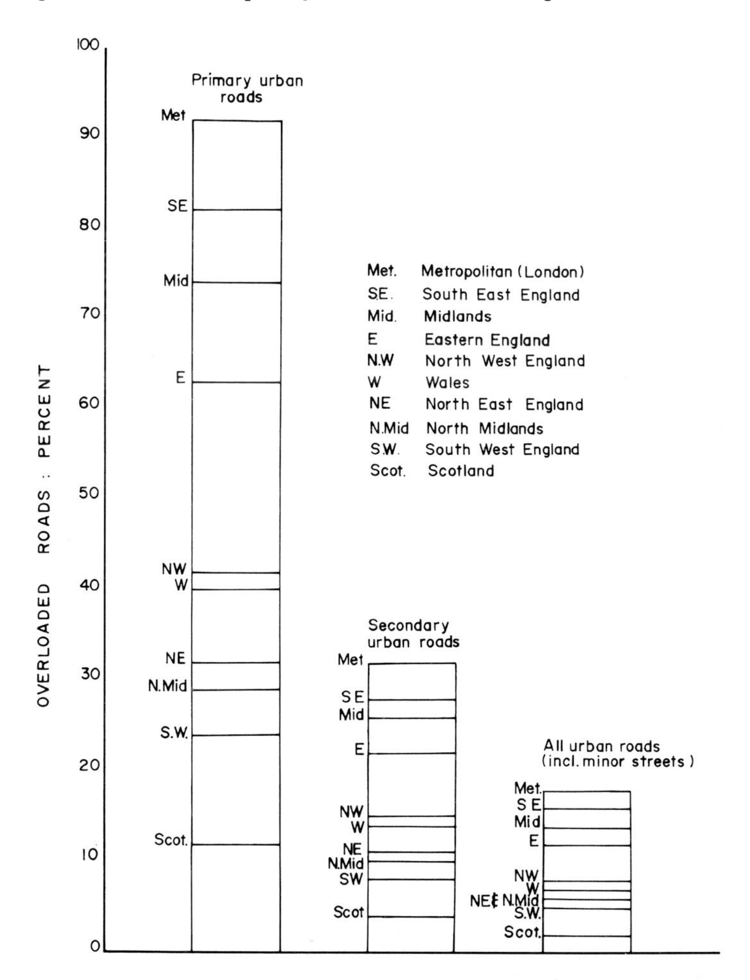

FIG. 2.4. Regional variation in overloaded roads, 1960: significantly overloaded urban roads occurred only in the southern and eastern parts of England in spite of the rapid growth in car ownership during the fifties.

intense concentration of white-collar jobs at the centre of the relatively more prosperous South-East Region. The rapidly growing commuter belt in the Home Counties enjoyed higher levels of car ownership than most of the country, and the consequences were to be readily seen on London's streets. After 1952, speed fell to barely more than 8 miles per hour in the peak hour so that vehicles spent nearly half their time on the central streets stationary rather than moving. By December 1958 the additional shock to the system of the motorized Christmas shopper brought the Capital's roads close to fulfilling, if only temporarily, that popular notion of traffic "grinding to a halt".

Summary and Conclusions

Until the early sixties it was difficult to point to any substantial road scheme built in British cities since the 1930s. Moreover, what little bits of road had materialized by that stage had slavishly followed archaic design principles worked out during the Second World War and published immediately afterwards in a Ministry design bulletin. This expressed the view that urban motorways would be needed only in exceptional circumstances. Throughout the fifties it was evident that the Ministry considered circumstances far from exceptional and on occasions had positively discouraged urban motorway schemes.

Fortunately for the Government, there were factors working in its favour. It transpired that urban streets had plenty of spare capacity, and factors such as the demise of the tramway systems added yet more. Another factor was that in most of the large industrial cities where jobs were highly concentrated, growth in car ownership lagged behind the rural and southern districts. Consequently, only a few places – London being a notable case – were feeling intense traffic pressures by the end of the fifties.

Why the Government discouraged urban motorway planning in the 1950s is a question not easily answered. It is tempting to point to Ministry of Transport forecasts of future traffic growth as a strong influence. These forecasts consistently and seriously underestimated the situation. With forecasts so wide of the mark, and with the accompanying notion that the ultimate level of ownership was but a small proportion of that later anticipated, it is quite possible that the Ministry simply thought there was no demand for additional road capacity on a scale implied by motorway designs. Indeed, according to one commentator[6] the Ministry gave many people the impression that it considered extra spending on urban roads as quite unnecessary; an impression supported by other evidence[7] which suggested that until the late 1950s the Ministry had seen the motor problem mainly in terms of the car's impact on the country districts.

It is difficult, however, to square such ignorance or complacency with information that the Government were receiving from other quarters. The Ministry's Chief Highways Engineer and the Director of Road Research in the Department of Scientific and Industrial Research were both "driven to the conclusion" by the mid fifties that a comprehensive solution to the urban traffic problem had to include new roads.[8] Moreover, in their opinion the greatest benefits were likely to come from building these roads as urban motorways. Therefore, one is inclined to see the Government's reluctance to endorse the advice of its experts (albeit experts with an interest in road building) as, at best, wishful thinking or, more likely, a conscious attempt not to detract from a policy already decided upon. This was to divert as much as possible of the limited funds made available for roads to fulfilment of its trunk roads policy.

But whatever the real reasons for the Ministry's cavalier treatment of urban roads and traffic in the fifties, by the end of the decade it left the incoming Minister, Ernest Marples, with few options and little time to do something positive. In the event, Marples grasped a policy designed, as he was to put it, "to extract the last ounce out of the existing street system".

CHAPTER 3

Squeezing More Out of Existing Streets

IT was recognized in Whitehall at the end of the fifties that with traffic growing rapidly and with few urban road schemes in the pipeline, traffic capacity, for a few years at least, would have to be squeezed increasingly out of existing streets. But this was no easy task. It depended on the skills of the traffic engineer and at this time such skills were in very short supply. In 1961, for example, less than a third of all counties and less than one-fifth of all county boroughs employed such persons. Therefore, there was need to make a little go far.

The solution hit upon was to use London as a national test bed, shop window and propagation centre for traffic engineering techniques. The basic ingredients in this cornucopia were threefold: first, a traffic problem that by common consent was Britain's worst; second, London's unique standing in possessing a traffic authority that, by historical accident, happened to be the Ministry of Transport; and third, a new Minister of Transport, Ernest Marples, with a remarkable talent for publicity.*

The Street Parking Problem

If more traffic was to be squeezed through London's streets, one of the first essentials was to tackle the on-street parking problem. During the 1950s, London's streets were filling very rapidly with parked cars. In 1951, a year after

*Ernest Marples was one of Britain's more colourful Ministers of Transport. He entered Parliament as Conservative MP for Wallasey in 1945 and moved through a number of minor ministerial positions until Harold Macmillan gave him an opportunity with the Post Office portfolio in 1957. This he made the most of, and in the ministerial reshuffle following the election victory of 1959 Macmillan transferred him to a Cabinet position as head of the Transport Ministry (by then separated from Civil Aviation). There he was to remain until Harold Wilson's Labour Government came to power in October 1964 – a length of stay at the Ministry surpassed since the Second World War only by Alfred Barnes. Possessing considerable energy, Marples went on to confirm Macmillan's initial expectations of him (Macmillan had written:[1] "Defence I have given to Watkinson and Transport to Marples. These are both 'self-made' men (*novi homines*) and will both do well"). During his stay at the Ministry he succeeded in setting in train a number of important transport investigations that were to produce famous reports, like Dr. Beeching's report on reshaping British Railways. Marples left Parliament in 1964, was made a Life Peer in 1974 and died in Monte Carlo in 1978.

petrol rationing ended, about a quarter of the usable length of street in the inner area was occupied by stationary vehicles. By 1955 this proportion had grown to at least 40 per cent over-all, and in some parts of Westminster exceeded 80 per cent. Parked vehicles in inner London now stretched for no less than eighty miles. Thus, with the combination of much less space available for the moving vehicle and more vehicles wishing to move, speeds were falling rapidly by the mid fifties.

By this stage, however, the shock waves had registered enough to induce Parliament to pass legislation permitting parking meters to ration out the kerb-side space. Meters as a solution to London's problems had been suggested on an

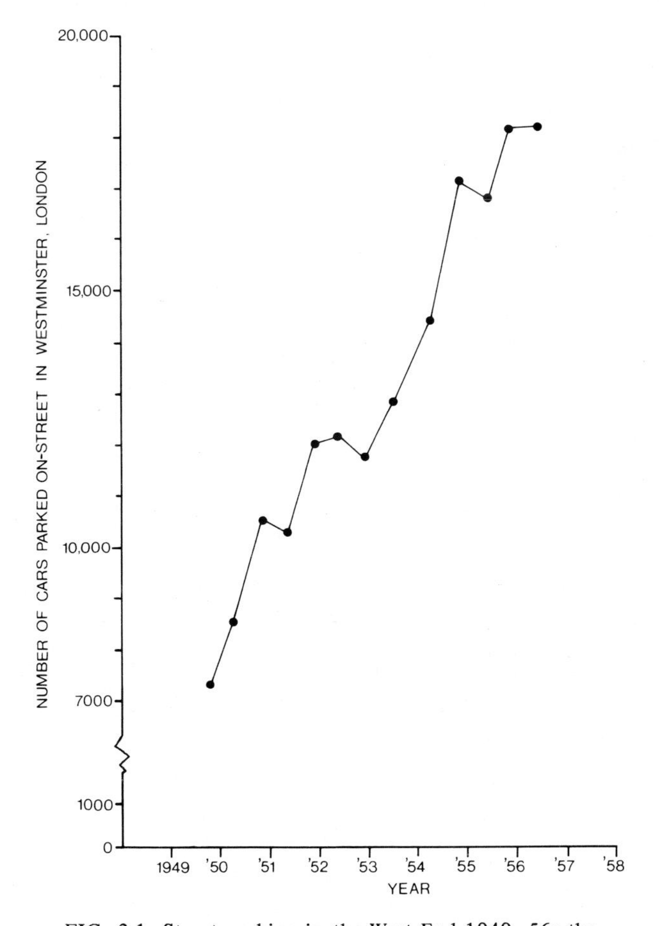

FIG. 3.1. Street parking in the West End 1949–56: the growth trend relates to cars parked on streets in the City of Westminster (excluding Royal Parks). The growth during the 1950s was dramatic and led to problems for the moving vehicle.

experimental basis by the London and Home Counties Traffic Advisory Committee* in the early fifties. But the motoring lobbies were strongly opposed to the idea and it was only after a bargain was struck between Boyd-Carpenter, (the Minister of Transport and Civil Aviation) and the motoring organizations, whereby net revenues from meters would be used to provide off-street parking facilities, that legislative provision was made. Meters were enabled by the 1956 Road Traffic Act.

Nevertheless, it wasn't until July 1958 (by which time parked vehicles filled around ninety miles of street) that this new form of street architecture finally made an appearance; and then only in small numbers. London's first scheme in the Mayfair district of Westminster for example produced but a few hundred meters and, overall, numbers were to remain quite small for some time. One reason for this was that, to be effective, meter schemes needed enforcement in the context of general regulations on parking and waiting. Here lay a problem. At this time enforcement of the parking meter regulations was in the hands of meter attendants. But, these attendants had no powers over unmetered space in the metered areas. Such powers rested solely with the police. Unfortunately, the police, short-staffed and under pressure from other quarters, particularly in London, were inclined to place waiting and parking regulations well down the list of priorities; their attention to such matters was frequently irregular and erratic. As a consequence, many motorists were to be found disregarding meters and taking their chance in unmetered spaces which brought the system as a whole into disrepute. Indeed, there were reports of the West End meter system collapsing soon after its introduction due to a lack of enforcement of the general regulations on parking and waiting.

This critical role of enforcement (and the advantages of streets cleared of parked cars) was emphasized at Christmas-tide 1959 by "an emergency plan"[3] known as the Pink Zone, launched with customary flair and publicity by Marples within weeks of his arrival at the Ministry. The Pink Zone did little more than try to make effective for a temporary period the then current parking and waiting regulations. It did this by enforcing more stringent controls over parking and waiting in the heart of the West End, and by offering a carrot to motorists in the form of extra parking space adjacent to the central shopping area. It was an attempt, in fact, to avoid the near-chaos produced the previous Christmas when extra cars were brought into the Capital by shoppers.

The initial reactions to the scheme were very favourable. The *Guardian* reported[4] that "all day the traffic flowed smoothly, at more than twice the

*The London and Home Counties Traffic Advisory Committee was set up in 1925 with wide membership, mostly from local government, to advise the Minister of Transport on matters in the London Traffic Area, roughly twenty-five miles in radius from the centre of the city. The Royal Commission on Local Government in London, 1957–60, commented on its ineffectualness[2] and it was abolished in 1962.

normal speed, along streets where the all-day parker and unloading vans were nowhere to be seen". But a week later the paper commented:[5]

> "congestion reappeared at the end of last week as more and more motorists decided that the chances of having their cars towed away by the police were slight enough to risk unlawful parking. Whether traffic can move freely or not on a given day depends more on the chances of good behaviour than on the regulations; where enough motorists are prepared to defy the regulations they break down because there cannot be enough police to enforce them at every spot."

Clear enforcement was a critical issue and a key to a successful on-street parking policy.

The breakthrough in this regard came the following year, when the 1960 Roads and Road Improvement Act extended the fixed penalty system for parking offences. More importantly, it introduced for the first time persons under the general direction of the police, known as traffic wardens. Wardens were able to discharge functions normally undertaken by the police in connexion with road traffic law including enforcement of general parking regulations. Now with such a specialized force commanding the necessary powers, the control of parking was potentially a much easier task.

It was still necessary to use these new powers effectively in the context of traffic management as a whole. What was needed was a broad plan for controlling on-street parking linked with other techniques for improving the flow of traffic. However, by the time the Roads and Road Improvement Act was on the statute book, steps to do this had been taken already.

The London Traffic Management Unit

In February 1960 the Minister announced that he was establishing within the Ministry a special unit to plan and put into effect schemes for improving London's traffic flows. It was to be under the direction of George Charlesworth, seconded for two years from the Road Research Laboratory. Its immediate tasks were threefold: first, to improve circulation on the major streets of Central London; second, to improve conditions on the radial and circumferential routes to and around the centre; and third, to deal comprehensively with street parking in the centre.

The fact that such a unit could be set up with such a remit owed a great deal to historical accident. The London Traffic Act of 1924 had placed all matters concerning traffic regulation in London in the hands of central government – elsewhere in the kingdom such powers rested, except in the case of trunk roads, solely with local authorities. The Minister did have a duty, however, to consult the Home Counties Advisory Committee, and this had had the effect of reducing the speed with which new regulations could be introduced. But this procedure also was streamlined when the 1960 Roads and Road Improvement Act reduced to a limited number of cases the need for the Minister to confer with the Committee. By the time this Act was on the statute book, the Unit had recruited most of its key personnel, and in the autumn of 1960 it began its work in earnest.

The Unit started in the centre of London and worked outwards. It had been known for some years that traffic flows could be eased and traffic speeded up by restricting streets to single-direction movements. For example, a large and

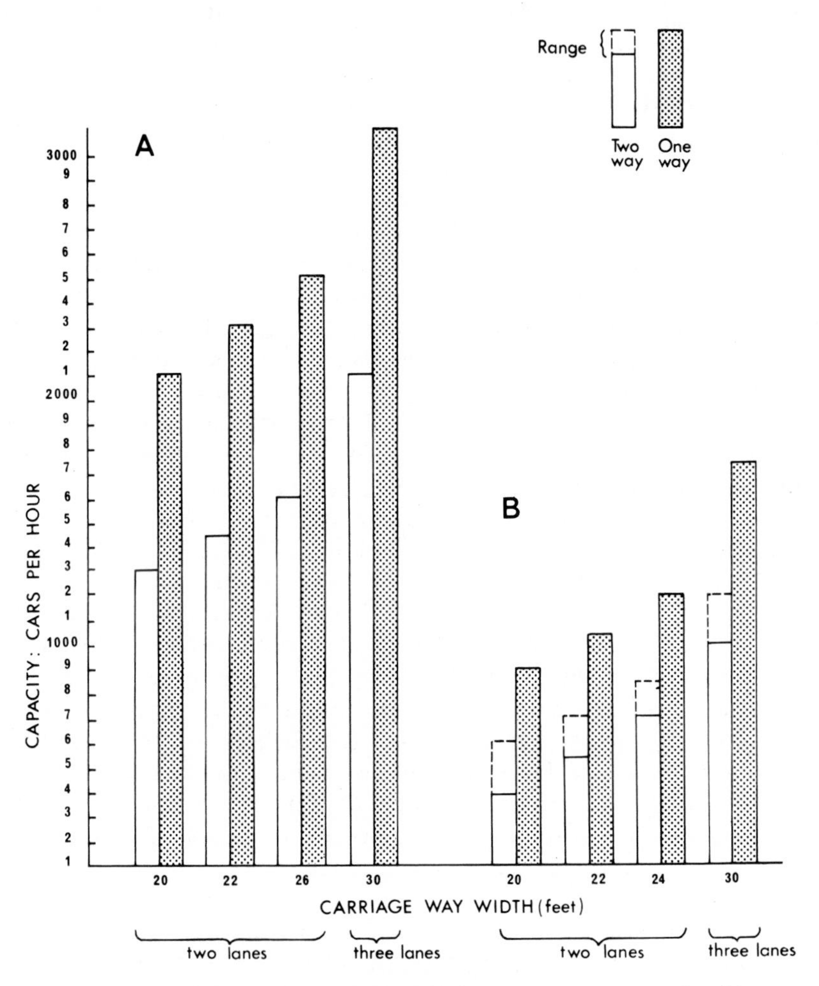

FIG. 3.2. Additional road capacity provided by one-way streets and waiting restrictions: (A) represents the situation for an all-purpose road with no frontage access or standing vehicles, and with negligible cross traffic; (B) portrays the situation for a similar all-purpose road but with capacity restricted by waiting vehicles and junctions. The additional capacity provided by one-way streets is considerable.

successful "cog and wheel" circulation system had been introduced in Central Birmingham in the 1930s. One-way systems were quite common also in London during the fifties but most had been fairly small and isolated. The Unit made an early impact here by introducing to Central London a number of major one-way schemes which were blended slowly into a general strategy. Notable in this regard was the Tottenham Court Road scheme which used Tottenham Court Road for north-bound and Gower Street for south-bound traffic, in conjunction with similar east–west movements based on New Oxford Street and St. Giles High Street. Other large-scale schemes followed, centred on Baker Street/ Gloucester Place and Piccadilly/Pall Mall. Slowly the Unit carried its work outwards from the Central area, so that by the beginning of 1965 there were in total forty-nine one-way schemes involving fifty-eight miles of street.

Although the idea of one-way streets had been around for a long time, in contrast, other measures introduced by the Unit included some novel developments. One such case was the introduction in 1964 of "tidal flow" schemes across a couple of two-laned Thames bridges (Albert and Hammersmith). The principle involved here was to switch capacity so that, at the busy time of the day, both lanes catered for the dominant traffic flow. In the case of Albert Bridge this meant two periods of the day when the bridge became one way, north bound in the morning peak and the reverse direction in the evening peak. The Hammersmith scheme operated only in the evening and involved switching the normally underutilized north-bound lane for the use of platoons of south-bound vehicles for a period of two minutes at a time.

Other new concepts also introduced as part of the Unit's work included yellow cross-hatched "box junctions", whereby drivers had to be reasonably certain of a clear exit before entering. Another was the introduction of urban clearways which imposed waiting and loading restrictions along significant stretches of major routes. A programme of clearways on twenty-one routes comprising some 120 miles of road was announced in 1964.

Meanwhile, the area covered by parking meters was being rapidly extended. In each of the two years after March 1960 the number of meters authorized for Central London doubled, and by 1962 they were beginning to appear for the first time in the outer areas. By March 1965 meter schemes were in operation or had been authorized for practically the whole of Central London (as well as in Croydon, Woolwich and Kingston-upon-Thames).

It wasn't all plain sailing, however. The Royal Automobile Club (RAC) in particular campaigned against the parking meter whilst a number of groups opposed various management schemes. The Knightsbridge–Chiswick peak hour clearway proposal of 1961, for example, brought considerable protest from traders and hoteliers situated along the route. London Transport also grumbled about the inconvenience caused by the need to re-route bus services. Although they acknowledged that bus services were faster and more reliable as a result of management measures, in some cases the degree of diversion from busy shopping

FIGS. 3.3 and 3.4. Before and after the introduction of traffic management: the photographs show Nags Head junction on the Holloway Road, London, before and after the junction was widened and traffic management measures introduced in 1965. The traffic management measures included turning the side road into a one-way street, prohibiting right turns and introducing clear lane markings.

FIG. 3.4.

areas was such that patronage was reputedly lost. The Piccadilly/Pall Mall scheme was a particular case in point. There was also "strong and determined opposition"[6] to some of the work of the Unit on the part of local authorities and some sections of the Ministry. The London County Council engineers, for example, were reported to have objected to the Tottenham Court Road scheme. Such a scheme ran counter to their proposals for building new junctions at St. Giles Circus and Euston Road, both of which were designed on the assumption of a widened, two-way Tottenham Court Road. But it wasn't always parties directly and adversely affected that adopted a critical stance. *Socialist Commentary,* albeit with a political axe to grind, also took exception in no uncertain terms to Mr. Marples' revolution, claiming that the quality of life had deteriorated for a majority of Londoners now that traffic was squeezed through formerly more or less peaceful side-streets.[7]

"Nothing wrong at all, thank you, officer—it's just that I'm a terribly slow reader." 2.iv.64

FIG. 3.5. Sir Obsert Lancaster's view of traffic management.

In reply, the traffic engineers, strongly backed by Marples, indicated what they considered to be the overwhelming advantages offered by their approach. In the metered areas of Central London the number of vehicles parked on the streets had decreased by 50 per cent, with the reputed consequence that journey speeds increased by 15 per cent. Considerable gains in traffic benefits and reductions in accidents were also claimed for the many traffic schemes. A study of six large one-way schemes in London, for example, estimated that there had been annual savings in traffic delays valued at £0·7 million for a capital (once and for all) outlay of £0·3 million. Examination of the accident records of fourteen London schemes over a six-month period before and after their introduction showed a 19 per cent reduction in all injury accidents and a reduction of 36 per cent in pedestrian accidents. Indeed, so pleased were the Ministry officials with their efforts, that a radical change in highway design was canvassed whereby urban motorways would be designed and operated as an overall one-way system.[8] But in terms of the over-all impact of their work, the traffic engineers pointed to a remarkable success: for the first time since the trams disappeared in the early fifties, the average speed of traffic on London's streets actually rose in both the peak and off-peak periods.

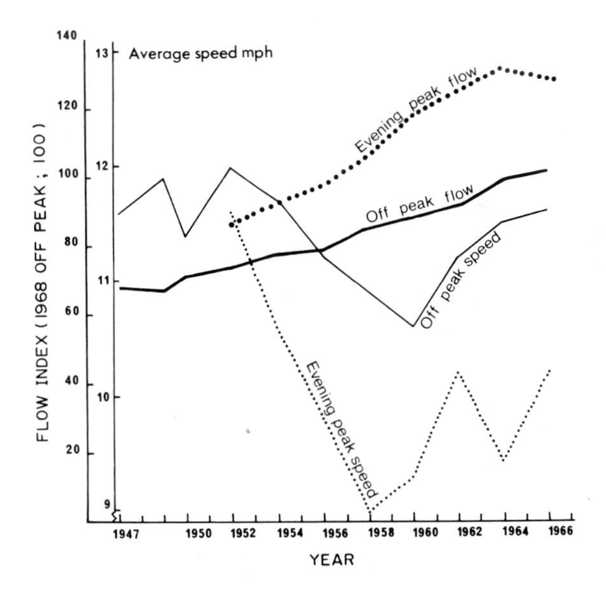

FIG. 3.6. London's traffic speeds 1947–66: after a continuous decline in the speed of London's traffic during most of the 1950s, the trend was reversed in the early 1960s largely on account of traffic management. The slight speed increase between 1950 and 1952 was the result of tram services being withdrawn.

Summary and Conclusions

Outside the towns in the early sixties the inter-urban motorway network was being pressed steadily forward, and the first few hundred miles of these high-capacity roads were helping to nullify the rising tide of vehicle ownership. In marked contrast, within towns, very few new roads had been built or even planned for. Here, additional street capacity in the near future would have to come in a different way. It would have to come from what Ernest Marples was to refer to as a "policy aiming to squeeze the maximum amount of traffic . . . through the existing streets".[9]

But this was no easy task. To do it effectively required the application of skills, knowledge and techniques of traffic management, and these, at the beginning of the sixties, were in short supply. There was, therefore, a major job of first discovering how effective management measures could be and then publicizing the resulting benefits. London was to be used for this purpose.

London as a test bed and shop window for comprehensive traffic management had two advantages. The first, somewhat mixed, advantage was that it contained what one commentator described as "rather more than one-sixth of the hard-core traffic problem of Britain".[10] The second was that in London, Whitehall was the traffic authority (and was to remain as such until the formation of the Greater London Council (GLC) in 1965). Linked to these two factors were the undoubted political and entrepreneurial skills of an ebullient Mr. Marples.

A few months before Marples' reign at the Ministry came to a close he was to spell out in Parliament what he saw as the success of his London policy of traffic management.[11] He pointed out that in 1959 the Labour Opposition had been very sceptical about whether the Tories could do anything with London's traffic. He then went on to point out that between 1958 and 1962 the traffic flow in Central London in the evening peak had increased 19 per cent, but journey speeds had increased 14 per cent whilst casualties, as measured over a slightly shorter three-year period, had decreased 6 per cent. These were statistics to which Marples gave general approval in the phrase, "if traffic moves faster and accidents go down that is what we want". However, some wanted, rather more particularly, quieter streets to live in, and in this respect Marples' single-mindedness was criticized. One outcome was that traffic management schemes implemented later in the decade were to show some consideration for related environmental issues. But in the meantime, Marples had shown in London what comprehensive traffic management could accomplish in terms of simply keeping traffic moving once the need had been established. It was to prove a lesson that the provincial towns and cities were to draw upon increasingly and successfully in the years to come.

CHAPTER 4

The Buchanan Report: Traffic in Towns

IN December 1959 during one of the first debates on transport after Marples had taken office, he canvassed the idea of a study group consisting of, amongst others, architects and town planners, to see whether it was possible to come to terms with the motor car.[1] Eighteen months later, having been impressed by Colin Buchanan's book, *Mixed Blessing,*[2] reputedly read on a trans-Atlantic flight, the Minister plucked its author from the relative obscurity of the Government planning inspectorate to lead a team "to study the long term development of roads and traffic in urban areas and their influence on the urban environment".*

The Report in Brief[4]

The initial focus of the team's work was to develop further Buchanan's earlier theme in *Mixed Blessing* of conflict between the beneficial aspects of the motor vehicle, especially "its ability to provide a door to door service" and its deleterious side effects of noise, visual intrusion and accidents (at the time there were over a third of a million road casualties each year, three-quarters of them in towns). Moreover, the team concluded that with little prospect of a serious competitor to the motor vehicle emerging in the near future, vehicle numbers were going to grow at an alarming rate and were expected to double within ten years. They added that "The population appears as intent upon owning cars as manufacturers are upon meeting the demand" and was unlikely to be deterred by congestion and frustration. The Report then warned that the potential increase in the number of vehicles was so great that unless something was done, conditions were bound to become serious within a comparatively short period of years. Either, as clearly stated in the Introduction to the Report, "the utility of vehicles in towns will decline rapidly, or the pleasantness and safety of

*However, Sir Richard Clarke[3] reported that one of the most important factors leading to the commissioning of the Buchanan Report was the deliberation of a joint study group of the Society of Motor Manufacturers and Traders, the Treasury and the Ministry of Transport set up to examine the implications of the great expansion of the car population upon which the motor industry's production plans were based.

surroundings will deteriorate catastrophically; in all probability both will happen together".

This conflict, it was felt, would not be alleviated by endorsing the "very powerful force towards a spread of development". To do so would merely invoke new problems at the expense of losing "the degree of compactness and proximity which seems to contribute so much to the variety and richness of urban life". On the other hand, the conflict could, it was thought, be greatly reduced, although not eliminated ("there is no straightforward or 'best' solution"), by attention to the *design* of compact towns and cities, by careful arrangement and disposition of buildings and activities to enable the motor vehicle to be used to its best advantage. Indeed, the Buchanan team saw their basic task as essentially one of exploring a problem of design, and this was a matter of rationalizing the arrangement of buildings and access-ways.

The basic design principle adopted was one of "circulation" and here the analogy of designing the internal arrangements of a large hospital was adopted, invoking ideas of corridors of circulation and areas of environment (representing wards, operating theatres, laboratories etc.). The city counterpart was referred to as "urban rooms", not free of traffic but designed in such a manner that their traffic was related in character and volume to the environmental conditions being sought. Large-scale application of the concept "results in the whole of the town taking on a cellular structure consisting of environmental areas set within an interlacing network of distributory highways".

An important point made was that "the function of the network would be to service the environmental areas and not vice versa", and from this axiom followed rules regarding the pattern, size and character of distributory highways. First, the pattern of the network must depend upon the disposition of the environmental areas, the kinds and quantities of traffic they generate and the associations that exist between areas (and, for these reasons, the frequent slavish adoption in past road plans of ring roads as a standardized pattern was frowned upon). Second, the network must be designed to suit the capacity of the environmental areas (here a water-pipe cistern analogy was used). And, third, it was necessary to introduce the idea of a "hierarchy" whereby the important distributors, referred to as "primaries" (with the function of canalizing longer movements in a fast, efficient manner) fed down through distributors of lesser importance to the minor roads which give access to buildings.

Whilst the team found these ideas concerning the network relatively easy to express, by comparison, the concept of environmental areas gave some difficulty. There was no connexion intended with the idea of "neighbourhoods" and its sociological ramification; they may be busy areas of any kind of development, but the important factor was that traffic within them would be subordinate to the environment. Here lay a problem because, as Buchanan and his colleagues noted, "so little serious research has been carried out into environmental standards". Nevertheless, it was presumed that the need for such standards was

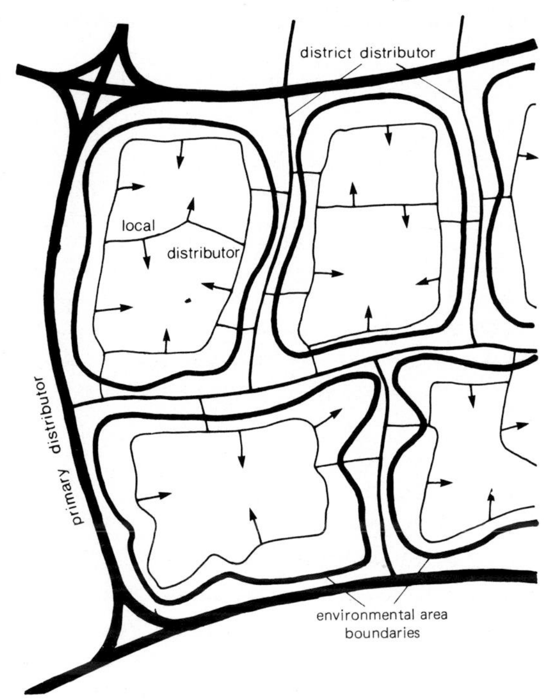

FIG. 4.1. The principle of a road hierarchy and environmental areas: these, combined with traffic architecture, formed the quintessence of Buchanan's approach to the design of towns for the motor age.

self-evident, and, indeed, was possible to determine. Once having agreed what these were, then it was possible to define the environmental capacity of streets, and, hence, of environmental areas too.

However, the capacity of an environmental area to accept vehicles, moving and stationary, depended also upon what Buchanan referred to as "traffic architecture" (a process whereby buildings and their access facilities were designed in a single process). By the skilful application of traffic architecture, more vehicles could be accommodated in an environmental area without violation of environmental standards. Not all the design concepts involved in this context were original – this was acknowledged – but what was new was a willingness to push them into the third dimension. When activities and movement were most intense and pressures at their greatest, then, the team concluded, the situation could be eased by complete redevelopment, by multilevel design and by placing the primary roads below decks providing access and pedestrian facilities.

In fact Buchanan viewed this as his distinctive contribution to the design problem. In an article published in 1961 he had written:

> "It might be objected that this [concept of environmental areas] is no more than the old precinct theory brought up again. In a sense it is, but . . . the precinct theory was always tied to the idea of areas of homogeneous land use *bounded* in the same horizontal plan by traffic arteries. On my definition the distributory network could as well pass underneath the environmental unit as round it."[5]

Such an approach, of course, placed a premium on an ability "to command the development or redevelopment of sizeable areas". Consequently, the presence of existing buildings which could not be redeveloped would "unquestionably affect the amount of traffic that can be planned for". The Report then added, somewhat prophetically:

> "unless the public accepts that there has to be comprehensive redevelopment over large areas, then the opportunities for dealing imaginatively with traffic will all be lost, and in the end this will severely restrict the use that can be made of motor vehicles in built-up areas."

These ideas were then examined in the context of four major case studies, with particular attention being given to how factors such as town size, density and the presence of historic buildings affected the application of the above "principles".

The team looked first at a small town, with the objective of demonstrating what was "involved in going the whole way with the motor vehicle". The town chosen for the purpose was Newbury (Berkshire), a market town of 30,000 people, with a dependent population in the surrounding area of a further 30,000.

The lesson of this study was that in a town the size of Newbury the potential increase of traffic would be far beyond that which could be accommodated by adapting the existing road system. Nevertheless, it was possible to provide for virtually all the vehicle use likely to be wanted, but it would require drastic and expensive measures on a scale hitherto unexpected for a town the size of Newbury.

The second study was of Leeds, a city then of about half a million people. Future land uses, movement patterns and modes of travel were again investigated and translated into vehicular journeys that could take place by the end of the century if there were no restrictions on such movement. The estimated (ultimate) daily total number of journeys in the city would be slightly over a million with 150,000 taking place during the peak hour, about three-quarters of the latter being work journeys by car. An attempt was made to devise a road system to accommodate this traffic, but it was concluded that there could be no possibility whatever of achieving this in practice. It would have required the complete redevelopment of some 2000 acres of the city centre.

An "intermediate" network was then examined in which 40 per cent of the potential demand for journeys by motor vehicle to the central employment zone, 70 per cent of demand for journeys to other inner employment zones and 100 per cent of journeys to outer zones could be accommodated. The build-up of the traffic resulting from those assumptions would still have required a substantial redevelopment of nearly half of the central business area to obtain satisfactory accessibility and a good environment.

Thus there was no escaping the need to consider curtailment or restraint of the full potential level of motor car use. Any necessary restraint was bound to fall first on the use of private cars for work journeys, and this was seen to have vitally important implications for public transport. But even with restraint, difficult and costly measures would still be required in order to ensure a good environment.

There are many cities in Great Britain whose old centres have historic and architectural character, the conservation of which prevents major reconstruction. Norwich, a cathedral city, then with a population of about 160,000, was the example that formed the basis of the third case study. The historic centre had many fine old buildings and the remains of the old city wall. The medieval street pattern was in itself an important part of the city's history; it was complementary to the buildings in scale, provided their setting and was essentially suited to the pedestrian. Therefore, the conclusion soon was reached that the environmental capacity of central Norwich was already being exceeded.

If the environment is held to be sacrosanct and if no major reconstruction could be undertaken, it followed that accessibility must be limited. The problem, then, was to establish the whole of the centre as an environmental area. This could be accomplished by locating the primary distributor road in the vicinity of the old city wall and by devising physical "barriers" to prevent cross-movement by through traffic. The old city would thereby be converted into four "rooms", each with its "doors" connecting to the new external corridor system (the primary distributor) along which the major car parks would be placed. But again the conclusion reached by the Buchanan team was that to achieve these satisfactory environmental standards, strict discipline of vehicle movement would be required.

The final case study was of 150 acres north of Oxford Street in Central London, a study which some later misconstrued as Buchanan's blueprint for Central London and all large cities. But again Buchanan was merely exploring design options. In this example, if all the potential vehicle use was to be provided for, the consequences would be networks impossibly large and complicated. Indeed, even assuming that the area was to be redeveloped completely and on several levels or decks, less than 30 per cent of what the team called the ultimate demand for the use of cars for the work journey could be provided for. With partial redevelopment (preserving buildings and areas of architectural or historical interest and large-scale motorway-type primary roads restricted to the edge of the area) the proportion dropped to some 20 per cent.

Thus, what these case studies showed was that accommodating high levels of motorization in cities at the same time as maintaining satisfactory environmental standards was going to be a prodigiously expensive policy. In fact, in the larger urban centres it was not just a matter of cost but of physical space also; it was physically impossible to cater for full motorization and, of necessity, some restraint on vehicle use would be required in these circumstances. Moreover, in so far as less elaborate schemes were desired "it follows inevitably that means must be found to restrict even further the usage of vehicles". This led back to what many would regard as the crucial argument in the Report, and what was to be referred to as Buchanan's "law". The axiom stated that if the quality of life in towns was to be maintained at a civilized level, then, as various case studies had shown, there were absolute limits to the amount of traffic that towns could accommodate. These limits would vary according to the size and density of towns, but, within both the absolute limits and a specified environmental standard, accessibility by the motor car could be increased by a readiness to accept and pay for the physical changes required.

The Report did not attempt to say at what level of traffic the choice should be made nor did it make positive recommendations as to the level of expenditure that should be incurred. But it did draw attention (as Wedgewood Benn, then

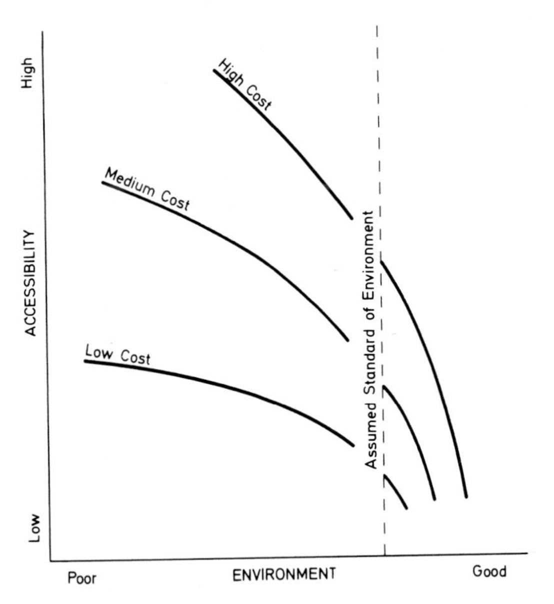

FIG. 4.2. Buchanan's Law: "within any urban area . . . the establishment of environmental standards automatically determines the accessibility but the latter can be increased according to the amount of money that can be spent on physical alterations".

opposition spokesman on transport, had done in a Parliamentary debate four years before[6]), to the ever increasing gap between private sector investment in vehicles and public investment in roads, and went on to argue that Society "will need to realize that it cannot go on investing *ad libitum* in motor vehicles without concurrently investing equivalent sums in physical accommodation for the vehicles". Nevertheless, the final choice, as the Report reminded its readers in its concluding paragraphs, was Society's. But, it was to be an informed choice which, in turn, required a comprehensive, quantitative analysis of traffic flows and an understanding of what the rate of traffic meant in environmental terms. The Report advocated, therefore, the preparation of "transportation plans" for this purpose, in order to supplement in large urban areas the statutory development plans required under post-war planning legislation.

Reactions and Responses

When the Report, *Traffic in Towns,* was published in November 1963, it immediately received popular (and ministerial) acclaim for its masterly, lucid survey of the situation. Buchanan's attempts to write so that the man-in-the-street could appreciate the problem paid dividends. At the time there was an intellectual vacuum to fill and the *Traffic in Towns* Report seemed to do this. Within four months 17,000 copies had been sold and Buchanan had received a "flood" of invitations to expand its thesis at conferences, meetings and gatherings throughout the country. Much of the ensuing debate appropriately focused upon what Buchanan himself saw as the prime issue; the commitment of resources to remoulding towns and cities for different levels of vehicle access.

The Report's own calculations were a starting point. Leeds, with half a million population, was estimated as needing a primary network of seventy miles (three-quarters to motorway standard) to meet about 40% of the estimated maximum demand for car use for central area work journeys at a cost of £90 million; in Newbury, a small town of 37,000 population, all the anticipated demand could be met, it was thought, with a primary network costing £4·5 million. Using these figures, an Oxford don, Christopher Foster, calculated[7] that the cost to central and local government of redeveloping towns and cities with a population of more than 20,000 to Buchanan standard (*sic*) would be about £12 billion or, including London, £18 billion. In addition, there would be the costs of private development. But, although the figures were put forward with great hesitation, they were nevertheless thought for several reasons to be an underestimate and that it might be necessary to find £22 billion which, if spread over forty years, would mean an annual charge of at least £550 million.

Support for figures of this order of magnitude came later in 1964 from Leicester where, under the guidance of its enterprising Chief Planning Officer, Smigielski, the city had produced a thirty-year traffic plan.[8] This plan was greatly influenced by the philosophy of the Buchanan Report – so much that it might be judged a further *Traffic in Towns* case study. Indeed, it did receive acclaim from one of Colin Buchanan's working party as being head-and-shoulders above previously prepared highway plans.[9] Nevertheless, Denis Crompton did question whether some local traffic flows were not going to be too high and the plan too much of a compromise, though, he added, it could not be called cheap. In fact to provide for only one-third of the estimated maximum demand in the central area of this city with a predicted 1995 population of 640,000, £135 million expenditure on "major structures" was proposed. Again, the very high cost of accomplishing reasonable access for the car without sacrifice of environmental conditions was evident.

However, not all observers were overwhelmed by the logic of these case studies nor by the arguments that massive investment was inevitable if the private car was to be catered for on a large scale. A more conservative and well-reasoned viewpoint was put forward in a powerful critique of the Buchanan Report, written in 1964 by the transport economist, Michael Beesley, and an American economist, John Kain.[10] The basic point made in the paper was that Buchanan had critically over-estimated the ownership and use of cars in cities in future years because the Report's analysis in, for example, the Leeds case study showed evidence of internal inconsistency.

The Report had argued from the basis of continuing and strongly concentrated employment in the centre of Leeds and only a moderate dispersal of the population, but with car ownership and use rising eventually to a level 25 per cent greater than the then current United States figure. The "internal inconsistency" was that all the available evidence pointed to the fact that only if Great Britain in future years resembled the United States in terms of all the factors determining car ownership and use, would the same levels of car ownership and use occur at the same levels of income. "By the Report's assumptions, these conditions are not satisfied in at least one crucial respect, namely in terms of the density of urban development." On the contrary, the densities assumed in the Report's case study of Leeds pointed to levels of ownership in twenty years' time no more than 40 per cent of the Report's estimate.

The crucial implication of this argument was that for the same levels of expenditure assumed in the Report a much greater proportion of the demand for car commuting could be accommodated or, alternatively, expenditure on highways need not be so great. But, the two authors also went on to argue that the Buchanan team underestimated the forces leading to the dispersal of cities and that "the latter is a natural corollary of the increase in accessibility postulated in *Traffic in Towns* . . . certainly if such suburbanization occurs, it will be accompanied by increased car ownership". But in these circumstances the solution of the problem of providing for the full use of the motor car is neither as difficult nor as costly as the Buchanan Report portrayed, because the cost of providing urban motorways depends very much on the intensity of development.

Thus, the Beesley and Kain thesis was that, either one way or another, schemes which the *Guardian* newspaper, for example, had referred to as being "so prodigiously expensive" would in fact not be needed to the extent assumed previously. Nevertheless, the Beesley and Kain argument merely reduced the scale of the increase in investment proposed; it did not dismiss the case for a substantial increase in expenditure and did not seek to argue otherwise.

The Government's response to this issue, as might be expected on matters of public expenditure, could be judged equivocal. Mr. Marples told Parliament in the major debate on *Traffic in Towns,* that the Government accepted the Buchanan Report in principle and that the modernizing and reshaping of towns should proceed within the framework of the main planning concepts embodied in the Report. He added that the Government accepted Buchanan's analysis and, therefore by implication, his assumption and arguments also. Amongst the more important of these was the case for keeping cities compact, or as Marples put the point, "keeping towns as towns" and the "country as country".[11] Marples also stressed the choices that all but small towns would have to make between traffic, environment and costs and added that "Each community would have to decide for itself". But to help local authorities in making these choices the Government was establishing (jointly between the Ministries of Transport, Housing and Local Government, and the Scottish Office) an Urban Planning Group to give advice and guidance, and in addition were encouraging new types of transport survey.

The latter, however, seemed to provide the Government with an opportunity for procrastination, or, as the Minister was to put the point, "We cannot do anything until we have had the results of these surveys".[12] The Minister was, however, perhaps doing himself an injustice. A commitment towards increased expenditure on urban roads was made which brought the retort from Professor Buchanan (quoted in a Roads Campaign Council Booklet[13]) that the figure seemed to bear little relationship either to the magnitude of the problem or to the rate at which the motor vehicle fleet was increasing.

Such comments did not appear to persuade the Ministry to be more generous with its anticipated provisions for the period after 1970. Using figures in the Ministry of Transport's Circular on Roads[14] (April 1965) as a basis, the technical panel to the Standing Conference on London Regional Planning advised the Conference[15] of a provision for the *quinquennium* after 1970 of £150 million for Greater London, a similar figure for the five other English conurbations and £190 million for other urban areas in England and Wales. Thus, a total expenditure of £490 million, or less than £100 million per annum, was planned for the early 1970s, whilst observers such as Foster and Reynolds[16] were suggesting that redevelopment of city roads along the lines illustrated in Buchanan's case studies required an annual amount perhaps five times this amount. Not surprisingly, the panel went on to remark "that in general the ideas put out in recent years for drastic re-casting of road networks in urban areas and for 'root and branch' improvements of town centres, were quite imcompatible with present and foreseeable financial provisions".

Summary and Conclusions

The Buchanan Report was without doubt the most fundamental and significant publication on the motor vehicle and its impact on modern society then published. Indeed, most would agree with the view that today it still remains pre-eminent, there having been few to rival its initial impact, its command of popular attention and its subsequent influence. And, like all documents that profoundly affect the course of events, it was to represent a watershed. The age prior to the publication of Buchanan's Report was one in which many refused to accept that car ownership would ever become widespread and general. The belief was also held that, eventually, once moderate resources were diverted from modernizing other transport sectors, the solution of more road capacity would catch up with the problem of traffic congestion. After Buchanan, there was a general acceptance that universal car ownership was a distinct possibility, and that achieving and maintaining a high degree of mobility in urban areas for the motor car was going to require a great deal of resources, regardless of whether towns remained "compact" or rapidly suburbanized. There were, of course, many, including the august members of the group steering the Buchanan Study, who opted for high levels of accessibility, the commitment of substantial resources and "nothing less than a gigantic programme of reconstruction".

From the standpoint of a later age these sentiments may appear curious and alien. But it is as well to remember that this was a time when the job of rebuilding cities blitzed during the war was still in hand and when the programme of developing "new towns" was in progress still. (The latter in particular was seen as a notable achievement which won international acclaim and it helped to maintain a belief, strong during the war and immediate post-war years, in a physical solution to urban problems.)

It was a time also when the birth rate was high enough to encourage thoughts of "new cities"; a time when the rate of economic growth was great enough to support ideas of comprehensively redeveloping the old cities; and a time when the nation was about to endorse a manifesto of national and regional planning spiced with the ingredients of a "White-Hot Technological Revolution". In these circumstances it was possible to see how large-scale urban road building would complete the picture. There was, in fact, a strong link as the group steering the Buchanan team had noted:

> "Though the impelling force behind [total reconstruction] would be the pressing need to reorganise our cities for the coming volume of motor traffic, it should be possible in many cases to draw an extra dividend in the replacement of slums or unworthy housing. Indeed it is possible that

> a vigorous programme of modernising our cities, conceived as a whole and carried on in the public eye would touch a chord of pride in the British people and help to give them that economic and spiritual lift of which they stand in need."

In the early 1960s it was, indeed, just possible. The vision of a city reshaped around the motor car was arguably a vision still in balance with the national ethos.

It was soon to prove, however, a vision not in balance with national resources. Within twelve months of the publication of *Traffic in Towns,* the country entered one of the all-too-frequent stop phases of the stop–go economic cycle so characteristic of the British economy in the fifties and sixties. Most of Harold Wilson's "White-Hot Technological Revolution" remained on the drawing board and the Government's response to calls for substantially increased levels of expenditure on urban roads, never more than lukewarm, cooled appreciably. Combined with policies favouring the status quo on the question of spreading the traffic load by deliberate dispersal of urban activities, circumstances were slowly but surely pushing the Government into confronting an issue that for some time had been hovering in the background – the issue of deliberate planned restraint on the use of cars in cities. The Buchanan Report had shown that the issue would have to be faced eventually. Now, with public spending once more under pressure and with the better opportunities for squeezing the last drop out of London's roads increasingly spent, the moment had at last arrived.

CHAPTER 5

The Road-Pricing Debate

IN the 1950s there was little prospect of bringing order to the growing chaos on city streets by imposing general restrictions on vehicle use. With the rationing and controls of wartime and the immediate post-war years fresh in the mind, it was unlikely that popular opinion would have countenanced such a solution so soon afterwards. There was, of course, the occasional call from the wilderness. Sir Herbert Manzoni, City Engineer of Birmingham, said in 1956 that a part-solution to the problem of congestion in cities was to keep out as much traffic as possible.[1] But a more typical sentiment of the time was to be seen in the response Manzoni's comment evoked from the distinguished Professor of Civil Engineering, Fisher Cassie. Fisher Cassie remarked that "from the Middle Ages . . . through the days of the Turnpike Trusts to the zebra crossing of today, restrictions to movement and to the development of the vehicle have proved to be of illusory value . . . the road must be adapted to suit the vehicle so that it can go freely where it requires to go".[2] With such views legion in the 1950s it is not surprising that Colin Buchanan was able to pass comment in 1958 that there was no current of opinion then discernible which favoured deliberate reduction of traffic for the sake of order and safety in the streets.[3]

Establishing a Case

By the sixties, attitudes were changing. Firstly, there was an intellectual response to the problem with concomitant developments of theory, addressing such questions as "What was an efficient level of traffic on city streets?"; "What was the best means of achieving this level?" and if pricing was to be used, "What prices should be charged to achieve it?"

For many years a number of economists had been arguing that the problem of traffic congestion was essentially an economic problem. The reason for this was that *urban* motorists frequently did not pay enough for their use of a scarce commodity: "road space". As a consequence:

> "it can be said that traffic flow will tend to stabilize at the level at which the costs of road use *to those who find it just worthwhile to join the traffic* equal the benefits obtained by them from their journeys. In other words, the level of traffic flow will depend on decisions taken by individuals

taking account of the costs and benefits *to them* associated with road use. But from the point of view of the traffic as a whole this is an unsatisfactory state of affairs, for the individual road user when making his decision does not – indeed he cannot – take into account the costs imposed by him on others. He assesses his private costs but ignores the road use, congestion and community costs. It follows that so long as the volume of traffic in conditions of congestion is determined by each road user considering only his own costs and benefits, traffic volumes will be larger, and costs higher, than is socially desirable."

"If we want traffic flows to be at levels most beneficial to traffic as a whole we must somehow make car users take into account the costs imposed by them on traffic as a whole, i.e., the congestion costs. One way of making road users 'take account' of the costs imposed by them on others is to make them pay those costs."

This quotation comes from a Penguin book, *Paying for Roads,*[4] published in 1967, but the basic philosophy was first mooted soon after the Great War of 1914–18. Not until the 1950s, however, did significant numbers begin to broadcast the concept. Stimulated no doubt by the rapidly deteriorating traffic conditions in towns and cities the idea soon began to catch on. Consequently, the beginning of the sixties saw a spate of papers advocating the introduction of road congestion pricing as an alternative to allowing congestion alone to ration out the limited road space available. Most came from academic economists and appeared for the most part in the professional literature.[5] However, there were notable exceptions. These included a "transport manifesto" developed under the auspices of *Socialist Commentary* and published in April 1963,[6] and an article by Gabriel Roth in a 1962 issue of the Conservative Party's *Crossbow.*[7]

This lobbying by the economists appeared to pay dividends when a panel was set up in 1962 to examine economic and technical issues of road pricing. The investigation was inspired not by the Minister but by Ministry of Transport staff, and it was notable that the Panel was neither appointed by nor reported to the Minister. Nevertheless, the fact that such a panel had been set up at all to examine such a subject under the auspices of the Ministry, was itself a remarkable development in transport policy.

The Committee was established under the Chairmanship of Reuben Smeed and for the most part was drawn from the ranks of the same economists who have previously publicized the cause of an improved pricing system for urban roads. But this committee composition reflected in large measure the nature of the terms of reference; the stress was not upon establishing a basic case, but upon *how* various forms of taxes could be levied so as to differentiate between more or less congested roads and upon the order-of-magnitude benefits each form would produce. Indeed the Committee was not required to consider the social and political aspects of the alternative charging methods, although, as they remarked, it was difficult to extricate completely the technical and economic issues from these (more delicate) matters. In fact at times the boundary was crossed and the panel "assumed" that the introduction of road pricing would be accompanied by a corresponding reduction in existing taxes so that the motoring population as a whole would pay no more than it would otherwise have done.

Of the different methods of charging considered, the Panel found little advantage from levying fuel tax at different rates in different areas. Significant advantage would come from a parking tax (in spite of its tendency to attract through-traffic to less congested city centres) and a system of daily licences for motoring in congested areas. But, best of all, would be the direct pricing systems using various forms of metering. Metering fell into two basic categories: either systems attached to the vehicle, analogous to sealed taxi-meters which worked continuously when the vehicle was in use, or systems which at various points along the side of the road would identify and monitor the movements of an individual vehicle and then transmit the information to central computers which would then calculate and compile the appropriate charges. These latter systems were analogous to telephone charging methods and were seen as exciting new technologies resulting from developments in eletronics made during the 1950s.

The chief conclusion of the Panel was that there was every possibility of developing the direct pricing technology into an efficient charging system that could yield substantial benefits on congested roads. In fact it was tentatively estimated that the measurable net gain to the community from higher speeds associated with fewer and more evenly spread vehicles on urban roads would be about £100 million–£150 million per year (or equivalent to about 15–20 per cent of total taxes then paid for the purchase and running of vehicles).

The Panel reported to the Ministry of Transport in June 1964 about two years after it had been set up.[8] In the meantime, important developments had taken place. Not least of these was the publication of the Buchanan and Crowther Reports. The former had, of course, illustrated that restraint policies were inevitable. But, as one commentator pointed out, its preoccupation with the wider issue, namely the form of urban areas, meant that discussion of how this necessary restraint might be implemented was given a distinctly lower order of priority.[9] Nevertheless, the opinion was proffered that the permit system was clumsy from an administrative viewpoint; the road-pricing solution was a possible solution only in the long-term; whilst the parking option was favoured as an immediately available measure. Comments by the Crowther Group on the issue were not distinguished for their incisiveness either. The Group recognized the need for some deliberate restraint, but continually qualified their discussion of methods by warning that a car-owning electorate would not stand for severe restriction, adding, somewhat fatuously, that black markets and corruption are the inevitable fruit of such attempts at prohibition.

Ernest Marples, however, was not afraid to express publicly the essential point. Or, as he said himself during the House of Commons debate on *Traffic in Towns,* "in large towns and cities and the great conurbations there is no prospect whatever of catering for all cars which might ultimately want to go into them. I cannot stress that enough".[10] He then went on to draw attention to a Gallup Poll purporting to show willingness on the part of motorists to accept restrictions if this eased congestion. Two days later in answer to a Parliamentary question he responded by saying that he accepted the conclusion of the Buchanan and Crowther Reports that sometime in the future some limitation of traffic in the larger cities would be unavoidable.[11]

The Report of the Smeed Committee was released in what was realized to be a run-up period to the October General Election of 1964. The Government had been elected in 1959 and an election had to be held by October 1964. With the Opposition ahead in the opinion polls, it was, for the Government, a time for caution. Possibly reflecting this, Marples' reaction on release of the *Road Pricing* Report was rather circumspect. He drew attention to the need to know more about the important social, town planning and administrative issues and announced that the possibility of further studies of these wide ranging problems was then being examined.

He further commented that he had not yet decided upon what action to take and then added in characteristically Marples' style: "But I shall not funk it – any more than I have funked anything else".[12] This much we shall never know. Labour won the election and now the issues fell into the lap of two new Ministers of Transport, first Tom Fraser and then from December 1965, Barbara Castle.

Road Pricing Under Labour

Regardless of what a Conservative Government might have done given more time, both Labour Ministers appeared willing to go along to a surprising degree with "esoteric" restraint measures during 1965 and 1966. Six months after the election, Fraser referred to the need for radical and novel measures designed to relate the demand for road space to the supply.[13] He announced that feasibility studies of a number of measures were taking place, including direct pricing of road use of the kind considered by the Smeed Committee. Two weeks later he was to elaborate.[14] From a study in progress he hoped to get answers to questions such as whether or not there should "be supplementary licences for driving in some areas at times of the day when roads are most congested?" He added that supplementary licences might have to be fairly expensive if they were to be effective and that it was "not an easy subject", but stated that there were immense benefits if what Professor Buchanan called "optional traffic" could be cleared from congested urban roads. By the time Mr. Fraser left the Ministry, however, the study he had earlier referred to was incomplete. But the running was taken up by his successor, Mrs. Castle. In October 1966 during a visit to New York, she reportedly surprised Ministry officials by stating her preference for pricing the use of roads according to the amount of congestion.[15] She added that "the technological hindrances are not important" and that the development of the "little black box" was under way.

It was quickly apparent that the Ministerial conversion to the economists' logic was not universally shared. Regardless of what Gallup Polls might reveal about popular attitudes to traffic restraint, there were lobbies decidedly against "radical and novel" measures. Mr. Du Cann, chairman of the Conservative party, choosing to ignore Mr. Marples' comments eighteen months previously, attacked the "anti-motorist" policies of the Labour Government.[16] The motoring and road lobbies followed suit. The RAC described the Ministry's attitude as "negative nonsense";[17] the Director-General of the rival Automobile Association (AA) referred to the "Canute-like attitude to the tide of traffic by threatening 'black box' motoring";[18] whilst Lord Chesham, Chairman of the British Roads Federation, representing a large number of firms and organizations with interests in road development, commented that "Road pricing is simply just another way of restricting the road user and we shall oppose it every way we can".[19]

However, quite apart from the reaction of the lobbies, by the time of Mrs. Castle's New York press conference, there were signs that the bureaucrats were disagreeing amongst themselves. In response to the reports from New York, they stressed that no decisions had been taken and that an important report of studies was still awaited and not expected until early in 1967. (At one stage it had been expected by the end of 1965.) When the report, *Better Use of Town Roads,*[20] finally emerged, it concluded that direct road pricing was potentially the most efficient means of restraint but that it would take several years to develop and put into effect. Supplementary licensing, however, was treated without enthusiasm; the study group was far from satisfied that it would be practicable and was doubtful that there was a big enough margin of advantage over comprehensive parking control.* It was this latter idea that the study strongly favoured as the most promising method of restraint in the short-term.

The study appeared to make the most of the difficulties of introducing direct road pricing and supplementary licensing, and not only that but also the timetable considered necessary for resolving them seemed to contradict earlier estimates.† But there were undoubtedly problems and risks in focusing too

*This conclusion did not appear to be in keeping with the evidence. Studies done for the Group by Michael Thomson had shown that the benefits of a tax on *all* parking space would be only half that of a daily, supplementary licence. It is unlikely that parking control as envisaged by the Group would give greater economic benefits than a parking tax, although it could be as effective in reducing traffic. However, as the group pointed out, the prime objective was to make the best use of road space and not to reduce traffic willy-nilly. The best use of road space is achieved when the users who value their use less than the costs they impose have been restrained. Parking control is not a good way of selecting these. Therefore there were indications that the group lost sight of their prime objective.

†The Smeed Committee Report had talked in terms of the necessary technical development by manufacturers taking about a year or eighteen months, whilst Mrs. Castle in her New York press conference had referred to development being well on the way. Now in

much on "radical and novel" methods. Indeed, as Mr. Fraser had pointed out (ironically on the very day that the Ministry's powers to control London traffic passed to the new Greater London Council) "a town council had to do it, it is not something we can impose".[21] Moreover, although he indicated that the Government would be willing to pay for such an experiment, it was not quite so simple. As the *Better Use of Town Roads* Report pointed out, experiments would require legislation. It also appeared that time was getting short, and many shared Professor Buchanan's remark made in the autumn of 1966 that "we are getting perilously near the point at which traffic volumes require positive holding in check pending the execution of major capital works".[22] Therefore, if local authorities were to be persuaded to do something quickly to restrain traffic, the most risk-averse course of action was to concentrate upon persuading them to tighten-up on the control of parking. In effect this was what the Government chose to do.

Summary and Conclusions

Growing pressures on urban road space as a result of rapid increases in vehicle ownership in the 1950s resulted in economists and traffic planners advocating changes in the way motorists and others were charged or taxed by the Government. The economists argued that the serious congestion in towns arose from inappropriate and inadequate means of charging motorists for the use of city streets, and that a more efficient use of road capacity would ensue from different methods of pricing. Eventually studies were initiated to examine such possibilities. The first of these, the Smeed panel on road pricing, established whilst Marples was Minister of Transport, considered the technical and economic case. It reported favourably on ideas such as pay-as-you-go motoring and suggested that quite large community benefits would result from the implementation of such a scheme.

The timing of the Smeed Report was both opportune and inopportune. It appeared only months after the Buchanan Report presented a convincing argument that the restraint of traffic would be necessary in the future in all but the smallest of cities: road pricing was, of course, one way in which restraint could be achieved. On the other hand, a General Election was imminent and the normally enterprising Ernest Marples was cautious in his comments on the Report. The change of Government, however, did not appear to kill the momentum that had developed. In less than a year of Smeed's reporting, a new study had been set up.

Although the emphasis of the new study was less upon improving the system

1967, almost three years after Smeed had reported, the *Better Use of Town Roads* Report was saying that from evidence given to the Smeed panel, it seemed likely that it would take two or three years to find out whether a workable system could be devised, implying that such work had yet to start.

of pricing urban roads and rather more upon means of traffic restraint, the prime objective remained very much that of the Smeed Panel, namely making better use, in an economic sense, of existing road space. However, it was not until 1967 that the study reported, and by this time there seemed marginally less enthusiasm amongst the bureaucrats. This was reflected in the Report. Direct road-pricing was still seen as potentially the most efficient means of restraint but, in comparison with the Smeed Panel, more stress was placed on the "very difficult technical problems" and its long-term nature. But more significant perhaps were the Report's comments that road pricing had implications for motor taxation policy.

Most road-pricing advocates had argued that direct pricing besides rationalizing the use of existing road space would also provide a policy for investment. This was an aspect that the Smeed Committee chose largely to ignore, but it had, nevertheless, commented that one of the advantages of road pricing was that it would provide guidance as to places where the need for road improvement was greatest. Other economists went further and the case for road pricing had been linked with the case for a self-financing road authority, with the implication that such road taxes should be ear-marked (or hypothecated) for spending on roads.

This, of course, was a controversial if not emotional issue that had ebbed and flowed for most of the century. The Treasury had been able to break the link between motor taxes and road spending prior to the war and was intent on the situation remaining that way. Thus, when Lord Chesham wrote to *The Times* and said that if Mrs. Castle "wishes to make her proposals more acceptable then road pricing must be part of a general overhaul of road administration and finance with the income passed on to a national roads board to further road development",[23] Whitehall must have reacted nervously. Indeed, Christopher Foster who had been a member of the Smeed Panel (but by 1966 was Director-General of Economic Planning at the Ministry of Transport), had already remarked that there was a potential danger in the belief that road pricing would release a lot more money for road building.[24] No doubt also such dangers had not escaped the notice of the Treasury representative on the *Better Use of Town Roads* study group.

CHAPTER 6

Bridging the Gap

A FUNDAMENTAL conclusion of the Buchanan Report was that there were physical limits to the amount of traffic that could be accommodated once the traditional, compact British town grew to a certain size. This critical size was never determined with precision. It would, in any case, vary with circumstances; historical centres, for example, having lower limits than well-worn industrial towns. On the whole though, in spite of this imprecision, the thesis was accepted; in the large town, public transport eventually would have to fill a gap.

But there was also another aspect to limiting traffic quite apart from the long-term physical considerations. In this case the considerations were short-term and stemmed from the belief in the mid sixties that a rapid growth in car ownership and use was going to take place in the years immediately ahead, before resources could be directed to comprehensively developing urban areas along the lines illustrated in *Traffic in Towns.* At this time there was still every intention of building extensive Buchanan-type primary roads and it was recognized that these "networks will call for massive investment in road construction".[1] It was recognized also that this would take time. Therefore, with car ownership expected to grow by 60 per cent between 1967 and 1974, the situation was believed to be "critical".

Consequently, the spotlight fell on restraining the use of the car in congested areas. As the *Transport Policy* White Paper of July 1966 put it, deliberate measures of traffic restraint were required to ensure that the volume of traffic entering congested areas was sensibly related to the capacity of the road system. It went on to draw the conclusion (confirmed a few months later by the Report on the *Better Use of Town Roads*) that "at present, a thoroughgoing parking policy is the best method of achieving this".

Parking: Changing Perspectives

Most authorities were familiar already with administering parking *restrictions.* For some years waiting and unloading bans had been an increasingly common sight in urban streets and their use had been greatly encouraged by Marples in his campaign of squeezing more capacity out of existing streets. The introduction of the parking meter increased the scope for restricting parking and, after its initial use in Mayfair, London, in 1958, it eventually spread to other parts of the Metropolis and then to the Provinces. By 1965, twelve cities* outside London had parking meters in operation. Again, however, like the waiting and unloading bans, meters were treated initially by the various authorities as a means of releasing more street capacity for the moving vehicle in the context of traffic management. They were not looked upon in the early years as a useful element in a broader strategy of restricting the total number of vehicles parking in city centres. Indeed, there was a legal impediment to using them in this manner. The provisions of the 1956 Road Traffic Act stipulated that surplus revenues were to be used for the express purpose of providing more off-street spaces. In practice, of course, after meeting expenses the surplus was often rather small – 27·5 per cent of gross revenues during the first five years in London – and the number of spaces added off-street were far fewer than those removed on-street.

However, if parking was to be controlled in the interests of a traffic restraint policy, a much broader approach incorporating off-street parking too, was required. The importance of this was evident from the results of social surveys conducted in the early sixties by the Government Social Survey (Table 6.1). Over-all, almost four in every five commuter cars were parked off-street, and, what was more, a fee for parking was paid in only 3 per cent of cases. The implication of these figures was that most parking space was being provided off-street, freely by employers or local authorities.

To a large extent this state of affairs was the consequence of policies that local government had adopted with Government encouragement in the fifties and sixties when it was thought that the correct policy to pursue was to provide "adequate" off-street parking space. In many town centres, local authorities encouraged commercial interests to provide space, or more commonly set about providing much of this space themselves. Often it took the form of multistorey

TABLE 6.1. *Car Journeys to Work and Parking Location*[2]

Where cars were parked which were driven to work on the sample day	Whereabouts of household				
	Greater London	Other conurbations	Other urban areas	Rural districts	All such cars
On the street (%)	34	19	17	10	21
Off the street (%)	66	81	83	90	79
	100	100	100	100	100
Sample size	187	482	223	124	1016

*Birmingham, Brighton, Bristol, Cambridge, Leeds, Liverpool, Luton, Maidstone, Manchester, Newcastle, Plymouth and Southend.

or underground car parks and these were increasingly seen as symbolic of a progressive town council. By April 1964, 10,000 spaces had been provided in car parks of this kind in the provincial towns and cities of England and Wales. A year later another 7000 spaces had been added in nineteen new multistorey or underground buildings and yet another fifteen (with the prospect of 6000 more spaces) were under construction. Moreover, council parking spaces were often provided free, or at nominal charges, on the grounds that this was in the best interests of the town. Although the legislation that gave local authorities the powers to provide parking space and acquire land and appoint staff for this purpose (in England and Wales it was the Public Health Act 1925) did enable authorities to levy a charge, there was no obligation for councils to do so or to cover costs.

To supplement this space available for public use, development control powers were also used requiring commerce and industry to provide in new developments for the parking of employee and visitors cars. In the 1950s, when these requirements were first established, a number of councils prescribed standards requiring a *minimum* number of spaces. This number varied according to the size of the building or its workforce. For example, by the early sixties the standards for offices ranged from the London County Council's one car space for every 2000 square feet, to Kent's requirement of one space for every 350–500 square feet (depending upon total size of the building). The result was, of course, that a large number of parking spaces were provided in town centres over which the local authority had subsequently no control. In Central London, for example, by 1962 about one-third of all parking spaces not associated with private residences were of this uncontrollable type.

It was these parking policies that, from the middle of the sixties onwards, the Government was seeking to change in the interests of a traffic restraint policy. In March 1965 the Government circulated a planning bulletin, *Parking in Town Centres,* to all local authorities.[3] This gave detailed advice on all aspects of the problem and urged councils to adopt comprehensive parking policies. It drew attention to the "disadvantages" of requiring parking space in new developments as a condition of planning permission and to the fact that parking policies could be used to limit congestion.

In February 1968 similar advice was again given but this time in more forthright terms. On this occasion the Minister of Transport asked all urban authorities of over 50,000 population and all traffic and parking authorities in the provincial conurbations to submit to the Ministry "Traffic and Transport Plans".[4] These were to show, in particular, how a balance between traffic and road capacity was to be achieved in the years immediately ahead. To help them with their task, the circular included a detailed hypothetical study of an imaginary town with a population of 125,000. One important conclusion of this "guide" was that if work journeys by car could be kept at their existing (1967) level, this would prevent flows in the central area rising by 10 per cent by 1974. To achieve this, some long-term parking would have to be priced out of the centre to the surrounding fringe. Indeed, in the circular it was expected that control of parking would be the "crucial instrument" of traffic restraint and authorities were told that it was the total amount of parking space and the way it was used that had to be related in a realistic fashion to the capacity of the town's road system.

Urban Traffic Flows and Speeds 1966–71

At least in London the response to the Ministry's pressures was encouraging. The GLC, as the strategic authority responsible, pursued two courses of action. First, in 1966, it adopted proposals for substantially extending the area of controlled on-street parking from eight to forty square miles with a view to deliberately restricting the long-term parker on an unprecedented scale. Second, in 1968, revised parking standards in new developments were approved. These specified a *maximum* number of spaces closely tied to an expected number of business callers and the like. At the time when both of these proposals were adopted by the GLC, the agreement of the individual London Boroughs was needed for their implementation. Therefore, progress was slow. But, it was steady progress, and by 1971 the tide appeared to have turned. Compared with 1966, the total number of spaces in Central London had declined (Table 6.2).

However, the situation with respect to the flow of traffic in the central area was rather different. Although the number of parking spaces in Central London had fallen by 1971, the decline was marginal and not enough to offset the greater use made of those spaces left (Table 6.2). Moreover, traffic passing through the central area was increasing. In fact increasing through-traffic had been a marked feature of central-area traffic during the sixties; the increase had been about a fifth in the 1968–71 period alone. Consequently, there was a rise in the overall flow during the peak period in the centre. Although it was a very small rise (less than 2 per cent between 1966 and 1971), nevertheless it suggested that in London at least, effective parking control was proving not quite the "crucial instrument" for restraining traffic that the Ministry of Transport had expected.

Outside London it was rather more difficult to judge the situation. In spite of the Ministry's cajolery, there was little hard evidence by the beginning of the seventies that the provincial cities and towns were actively pursuing an explicit policy of traffic restraint. What Ministry of Transport statistics did show, however, was that, in cities both large and small, the flow of cars during the journey-to-work peak increased between 1967 and 1971 (Table 6.3). In fact in some cases the increases were quite staggering. Amongst the urban areas sampled* (these ranged in size of population from 75,000 to half a million) private car traffic in the central area increased on average by not a great deal less than *twice*

*Sheffield, Bristol, Leicester, Luton, Reading, Preston, Watford and Chesterfield.

TABLE 6.2. *Changes in Parking Supply and Occupancy, Central* London 1966–71*†

Type of parking	No. of spaces	% of total spaces	Maximum occupancy (%)	Maximum no. of parkers
1966				
On-street: free	30,000	24·0	87·4	26,208
On-street: metered	14,500	11·5	83·9	12,172
Off-street: public	25,500	20·5	66·9	17,061
Off-street: private	55,000	44·0	72·6	39,917
Total	125,000	100·0	76·3	95,358
1971				
On-street: free‡	13,000	11·2	82·7	10,750
On-street: metered	21,000	18·1	89·0	18,700
Off-street: public	32,000	27·6	76·0	24,320
Off-street: private	50,000	43·1	86·0	43,000
Total	116,000	100·0	83·4	96,770

*Central Statistical Area (approximately 10 square miles).
†From information supplied by the GLC.
‡This includes 8000 spaces for residents.

TABLE 6.3. *Urban Car Traffic: Percentage Change* 1967–71[6]

	Whole town	Central area
Peak		
Conurbations	+17	+16
Towns	+18	+29
Off-peak		
Conurbations	+18	+10
Towns	+14	+16

the national growth in cars licensed during the corresponding period. And, even in the centre of the five provincial conurbations for which data was available (Birmingham, Liverpool, Manchester, Leeds and Newcastle-upon-Tyne), the rate of increase kept pace with the 17 per cent national growth in licensed cars.

Of course, what is not clear from the data is whether, as in London, the effectiveness of any restraint on parking was being offset by increases in traffic passing through the city centre or through the urban area as a whole. In the case of the major conurbations, however, it is evident that this was not the case. For these localities, statistics from the national census (summarized in Table 6.4) show that, in all cases, car-commuting to the core increased between 1966 and 1971. In the widely varying conurbation centres of Birmingham, Manchester

TABLE 6.4. *Journey to Work by Private Transport in Conurbation Centres, 1966–71*

	1966		1971		1966–71
	Number	% of all work Journeys	Number	% of all work Journeys	% increase
West Midlands	27,137	25	34,524	33	27
South-East Lancashire	29,324	21	36,711	30	25
Merseyside	23,528	17	24,845	27	6
Tyneside	15,320	20	19,113	29	25
Clydeside	16,439	13	19,416	18	18

and Newcastle, this increase was as much as one-quarter. If so little had been accomplished in terms of effective parking restraint in these large cities, where the pressure was expected to be greatest of all, it is unlikely that the picture was very different elsewhere.

Thus, by the beginning of the seventies, restraint measures had either been neglected or applied to little effect. By all accounts, therefore, the beginning of the new decade should have witnessed thrombosis of the urban traffic arteries. Traffic, as the popular notion had it, should by then have ground to a halt. But it did not happen. In fact, quite to the contrary, congestion as measured by average traffic speeds, appeared to have fallen slightly in both the peak and off-peak in the central areas of town and cities (Table 6.5).

Why the serious situation, expected in the absence of ameliorating action by councils, failed to materialize was explained by a number of events. In the main, the expansion of road capacity was probably greater than had been foreseen. There were, of course, additions to this capacity as bits of inner ring road, started in the early part of the sixties, were completed in places like Coventry, Leeds, Derby, Reading and Preston. But, in the provinces particularly, useful mileage was obtained from large-scale traffic-management schemes with the spread of the valuable experience gained in London during the early sixties.

TABLE 6.5. *Trends in Average Speeds in Towns and Cities* 1967–71[7]

		Average speeds (mph)		
		London	Provincial conurbations	Provincial towns
Central peak area	1967/8	12·7	9·3	11·4
	1971	12·9	11·1	12·3
Off Peak	1967/8	12·1	11·0	12·3
	1971	12·6	12·6	14·6

To help the process along, in 1965 the Ministry of Transport had set up a Traffic Advisory Unit. At the press conference announcing the launching of the Unit, there were strong hints by the Minister that councils had not been doing enough in this direction, and it was stressed that traffic management was to be a "permanent technique" and an "integral part of town planning". With further encouragement and help from the Ministry a number of towns soon introduced comprehensive schemes. Reading, Berkshire, provided a good example. The backbone of the Reading scheme was a system of complementary one-way routes, north–south and east–west covering some ten miles of road in the town centre. The scheme was brought into operation in stages, starting in June

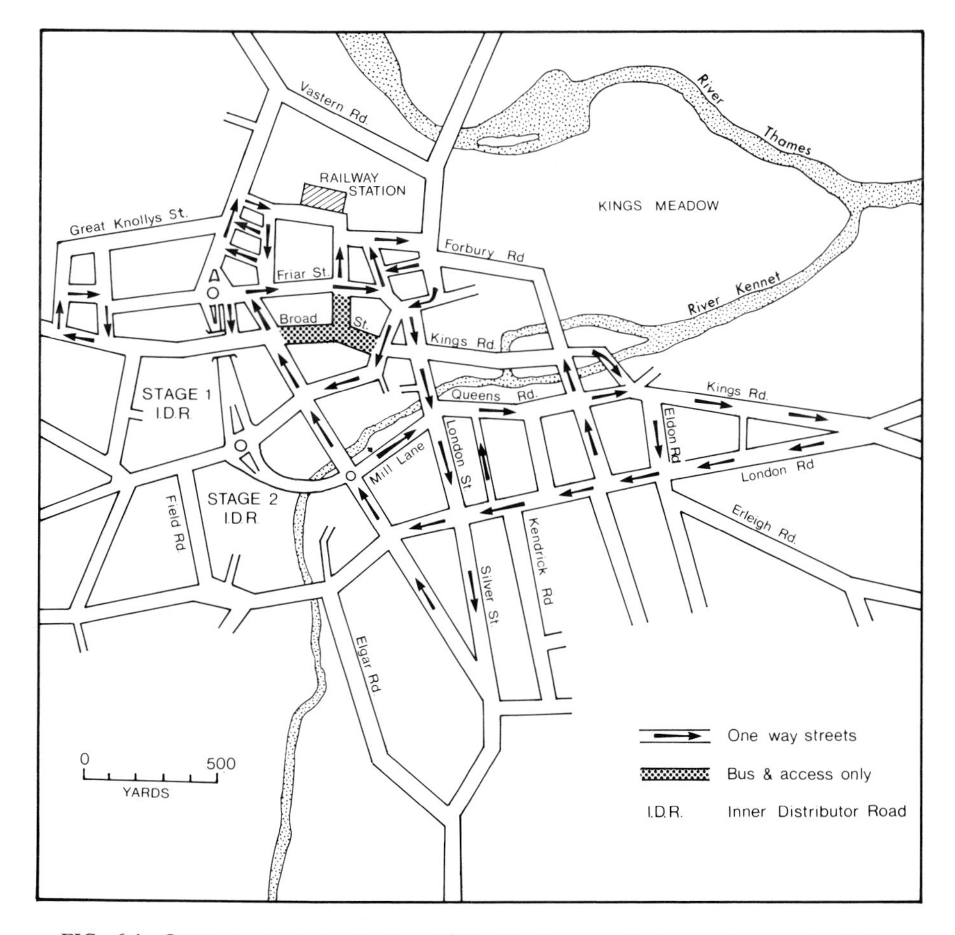

FIG. 6.1. One-way streets in Reading 1970: Reading provided a good example of the application of comprehensive traffic management in provincial towns in the late 1960s. Shown are those streets in November 1970 restricted to one-way working for general traffic.

1968 and finishing in November 1970, and quickly provided worthwhile results. Traffic flow greatly improved and average journey times on main routes decreased by a third or more in spite of there being no evidence of a reduction in the total number of vehicles entering the town. But, as in London, there was again some evidence of through-traffic increasing at the expense of local traffic.

However, if the expansion of capacity due to new inner ring roads and traffic management schemes was greater than expected, pressures on this capacity were probably less than anticipated. With the economy lagging, the growth in cars owned was not as fast as predicted; about half as fast in fact. As far as most big cities were concerned, there were far fewer central-area jobs to attract the car commuter. Industry in particular was leaving the centre of cities, either for suburban industrial estates or for smaller towns. In Merseyside, for example, the exodus was enormous, with one-third of all jobs leaving the centre between 1966 and 1971 (Table 6.6). This had two consequences. It meant that competition by motorists for driving-room in city centres was less intense than it would otherwise have been. Coincidence alone was unlikely to explain that the more the workforce had shrunk in the centre of the conurbations, the smaller the percentage increase in car commuting. (Compare for example the last columns of Tables 6.4 and 6.6.) It meant also that there were fewer lorries on the inner-city roads thus releasing more space for cars.

The full implications of the latter were revealed in a study *Cars for Cities* published in 1967.[8] The study had been set up by Marples in 1964 really as a complementary inquiry to Buchanan's inquiry.* As Marples said at the time:[9]

> "just as towns of the future must be rebuilt to come to terms with the motor vehicle, so the motor vehicles must be designed to come to terms with those towns. For example, can't we design vehicles whose size, power and manoeuvrability make them more suitable for town use?"

TABLE 6.6. *Changes in the Workforce of Conurbation Centres* 1966–71

	1966	1971	Difference 1966–71	% change
West Midlands	108,550	103,990	–4560	–4·2
South-East Lancashire	139,640	122,780	–16,860	–12·1
Merseyside	138,400	91,680	–46,720	–33·8
Tyneside	76,600	66,830	–9770	–12·8
Clydeside	126,460	110,950	–15,510	–14·0

*The Chairman was Lord Kings Norton and other members of the steering group included: George Harriman, Chairman and Managing Director, British Motor Corporation; Sir Patrick Hennessy, Chaiman Ford Motor company; Sir William Lyons, Chairman of Jaguar Cars; Sir Geoffrey Crowther, Chairman of the group that steered the Buchanan study; and Alec Durie, Managing Director of Shell Mex and BP Ltd and Director-General designate of the AA.

The answer to this and related questions was provided by the study report. Its conclusions were that better acceleration could help to increase road capacity, but on the question of size, the potential gains from smaller cars was less than often imagined. Even at moderate speeds, most space required by a vehicle was to provide a "headway" between itself and the vehicle in front so that it could stop safely. Knocking a few inches or feet off a car had surprisingly little impact on this "headway". However, the most important finding of the study was that uniformity of size of vehicle in the traffic stream was of great significance. It was much more important that vehicles were of similar size rather than the "average" vehicle being simply smaller.

In this context significant trends were taking place in the physical characteristics of cars. Although the *Cars for Cities* group did not examine these trends, new cars were, in fact, getting more uniform in length, more powerful and capable of accelerating faster from traffic lights and other hold-ups. By 1974, for example, the "average" new car registered was an astonishing 14 miles per hour faster in top speed than its 1961 predecessor.[10] In terms of length, although the average new car "grew" by just over three inches over the 1961–74 period, the important difference between smaller and larger cars decreased by a foot.* Therefore, without intervention or compulsion on the part of the government, the hidden hand of economic forces was slowly but surely moulding cars for congested city streets.

Summary and Conclusion

In the middle of the sixties, it was felt that a crisis of urban traffic congestion was imminent. Car ownership and traffic were forecast to grow very rapidly in the years ahead and, in towns, it was felt that traffic seeking to enter and circulate would inevitably grow faster than the capacity of the road system. Therefore, if the situation was not to get out of hand completely, "positive policies and strong local action"[11] were needed. It was expected that this would take the form of more rigorous comprehensive parking policies.

As it transpired, such positive, firm action failed to materialize; London being one of the few, albeit very important, exceptions. But even in London, on account of parking controls doing nothing to deter journeys passing through, traffic, as elsewhere, continued to increase. Flows of cars in more than a dozen places for which information was available increased considerably at a rate not distinctly different from the national growth of cars owned.

However, instead of speeds slumping under the burden of still denser traffic, the quite unexpected happened. Speeds crept upwards, even during the peak hours and even in the centre of the largest cities. New roads, of course, did make

*This is based upon the difference between the average for the largest 15 per cent registered and the average for the shortest 15 per cent.

a useful contribution to this relief. But it was from a more efficient use of existing roads and from changes in the character of the traffic that the real gains came. The skills and techniques of the traffic engineer were diffusing through the provinces, and, with industry moving to the edge of towns, there were fewer lumbering lorries competing with the car. The car, in any case, was getting faster and more manoeuvrable and more uniform in size, and was, therefore, making better use of the space left for it.

Thus, by the beginning of the seventies the doom-laden predictions of the mid-sixties had yet to materialize. Traffic had not been restrained to any appreciable degree by planned controls, and both flows *and* speeds were on the increase. With the prospects also of more spending on urban roads just around the corner, not surprisingly local authorities showed little sense of urgency in meeting the Ministry request for Traffic and Transport Plans. These short-term plans, illustrating the restraint methods councils were to adopt, were required originally by the end of 1969. But by early 1972 little more than a third of authorities had submitted completed proposals. Instead, at the turn of the decade, it was the plans for urban motorways and other large-scale road schemes that were capturing both attention and headlines.

CHAPTER 7

"Scientific" Road Plans and Urban Renewal

INCREASINGLY in the late 1960s the Government spelt out the need for traffic restraint in the larger towns and cities. However, it would be wrong to conclude from this that the administration was turning against ideas of building new roads in towns on anything but a substantial scale. On the contrary, the Government aimed to build primary networks in urban areas as soon as resources allowed.[1] Their basic strategy was first to fulfil the long-standing commitment to a 1000-mile programme of inter-urban motorways, and then to divert an increasing share of a rapidly rising roads budget to towns and cities. It was expected, in fact, that half the roads budget would be spent on urban roads by 1969/70 in contrast to approximately one-third in 1964/65.[2,3]

But, in the meantime, the parsimonious treatment of urban roads was to be turned into a virtue. As Ernest Marples told the Commons when debating the Buchanan Report in 1964,[4] each community would have to decide for itself where it stood in relation to Buchanan's law: the trade-off between accessibility and the environment. The choice needed to be based on proper factual information and the key to this was "the new type of transport surveys". These of course took time, but the initial channeling of funds into trunk roads provided the breathing space necessary to allow these surveys to be carried out.

The new type of transport survey to which Mr. Marples was referring was the land-use/transportation study. These assumed considerable significance during the 1960s, and they became, in fact, a major instrument of Ministry policy[5] – launched as they were on a wave of planning zeal that had an effect on all political persuasions.* The whole point of the planning exercise was economic growth, or, more precisely, an accelerated rate of growth, the fulfilment of which was fast becoming a national obsession. Allied with this obsession were ambitious plans for public spending. Public expenditure in 1964/65, as Keegan and Pennant[6] pointed out, was deliberately set to grow at least as fast as the

*The Tory Party, in the twilight of their thirteen-year post-war administration, formed the National Economic Development Council (NEDC), launched the Public Expenditure Survey Committee (PESC) System and the first of a series of regional planning studies. The Labour Government, on taking office in 1964, both intensified and formalized this process by forming ten Regional Economic Councils responsible to a new Department of Economic Affairs, which, in turn, spawned the 1965 National Plan.

Gross Domestic Product (GDP), and the assumption about GDP itself was a too optimistic 4·5 per cent. But, in the view of many people, more public expenditure was the *sine qua non* of growth. Growth required modernization on a national scale, and modernization meant more public spending on science and technology, on a revitalized regional infrastructure, and, by no means least, on clearing the slums and reshaping the cities around the motor car. The new, "scientific" land-use/transportation study, heavily geared to computer analysis and the strategic view of the city, fitted this scene perfectly.

Urban Motorways: The First Plans

At the time that Ernest Marples was calling upon local authorities to undertake the new type of land-use/transportation survey, a large number of road plans were already pouring forth from the Development Plan review process. The Planning Act of 1947, which had established an elaborate system of planning controls and procedures, required authorities to produce Development Plans for their areas. Many such plans had been produced and approved in the 1950s and these had to be reviewed and up-dated every five years. The quinquennial reviews were now taking place against a backdrop of rapidly rising vehicle ownership. As a consequence, the proposals for road widening and junction improvements, which were a feature of the 1950 Development Plans, were replaced in the early/mid 1960s with proposals for large radial and inner-city ring roads, often of motorway standard.[7]

Another marked and consistent feature of the early sixties planning scene was, of course, slum clearance and urban renewal, and attempts to combine road construction and redevelopment also figured strongly in the Development Plan Reviews of the 1960s. Indeed, there was every encouragement for authorities to approach the issue in this way. Sir Keith Joseph, when Minister of Housing and Local Government in 1964, had drawn attention to the large number of slums and areas of obsolescent housing in Britain, and commented that "urban renewal is wider than tackling the growth of traffic, but both can, to some extent, and in some cases, with the right policies be tackled simultaneously"[8].* Both the cases and the extent to which urban renewal and new road building could be tackled simultaneously proved considerable, and, in the documented plans for new urban roads, frequent reference was made to the opportunities that redevelopment provided. One example came from Sir Keith's constituency of Leeds, where a joint Ministry and City Council approach to urban traffic problems noted that

> "the need for extensive renewal at the heart of the city and in the area

*Interestingly, in 1961 Colin Buchanan, then a Ministry of Transport bureaucrat, had delivered a conference paper on the opportunities for traffic provided by comprehensive redevelopment.[9]

> immediately surrounding it thus offers a two-fold opportunity – for building anew in a newly created environment . . . and for using some of the cleared land to satisfy the extensive demand on land of an up-to-date system of highways and communications."

Here, at least, the opportunity was taken when one of Britain's first urban motorways, the Leeds Inner-City Ring, was built along the fringe of a large area of redevelopment to the north of the city centre.

Although opportunities for redevelopment influenced the precise location of many urban motorway proposals, their general form and scale were determined in a number of cases by quite extensive surveys of traffic. After an era of disinterest, and even discouragement, the Government's attitude to traffic and transport surveys was to change dramatically. As a consequence, in the late 1950s, the Ministry of Transport began to encourage highway authorities in the

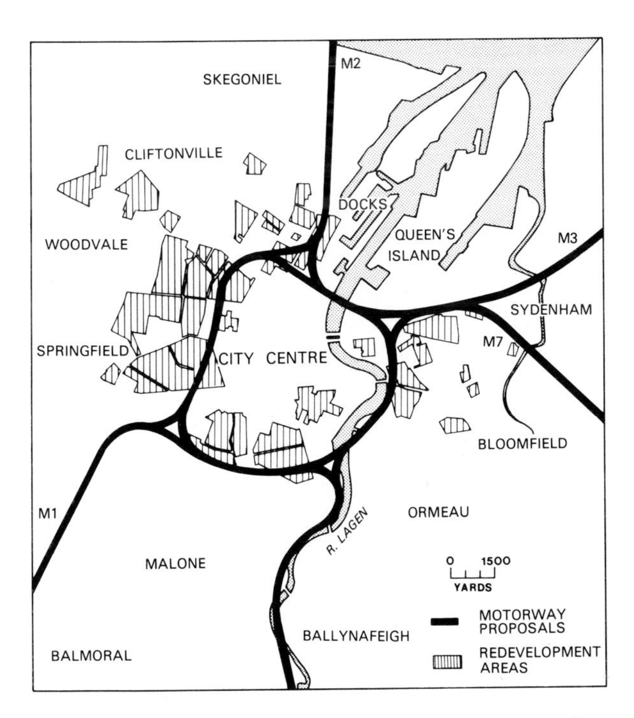

FIG. 7.1. Motorway proposals in central Belfast and redevelopment areas: "the opportunity to construct a ring road was improved by the 1956 Housing Act which made it possible to designate large central areas of the city for redevelopment". The road line shown is a modification of an early sixties proposal. Approximately two-thirds of the houses required for the ring road were in redevelopment areas.

major conurbations to come together to produce co-ordinated long-term highway plans for their areas. (At the time, of course, the conurbations suffered a fragmented local government system and continued to do so until the reforms introduced in 1974.)

The approach adopted was remarkably simple in concept, although fairly demanding of resources. In essence it involved surveys of existing traffic flows (including asking drivers where they had come from and where they were going), which were then projected forward a number of years on the basis of population and car ownership forecasts for different parts of the area studied. Road networks were then designed so that the predicted future levels of traffic would flow without undue congestion. Large-scale examples of these studies were the Highway Plan for South-East Lancashire–North-East Cheshire (SELNEC) produced in 1962, the Merseyside Conurbation Traffic Study and the Glasgow Highway Plan, both of 1965.

Land-Use/Transportation Studies

Buchanan and his colleagues, however, called for an approach to the problem that stressed a different emphasis. *Traffic in Towns* had highlighted the fact that street traffic was a result of the activities that took place in urban areas – traffic a function of land-use was the phrase coined. As well as a proper understanding of future movements as the basis for any plan, detailed knowledge of these activities and how they were likely to change was also required. It was this inter-relationship which was the prime focus of the land-use/transportation study which Buchanan was to recommend as the basis for future planning.

His sentiments were quickly endorsed by the Government, which expressed concern that the immediate reaction of many smaller local authorities to the Buchanan Report had been to dust off old plans and demand more money.[10] In a joint Ministry of Housing and Transport Circular sent to local authorities at the beginning of 1964, local authorities were called upon to prepare land-use/transportation studies.[11] Subsequently, to push councils in this direction, the Housing Ministry began to reject road plans in the quinquennial reviews of development plans where these reviews were not carried out in accordance with the Buchanan philosophy.[12] The Ministry of Transport, however, provided a carrot by offering help with the technical direction of the studies, and, perhaps more significantly, offering to share study costs with the local authorities concerned.

In London, the Ministry, jointly with the London County Council (LCC), was already supporting the London Traffic Survey. Initially this had the limited objective of collating simple facts about journeys, but soon it evolved to take on the more comprehensive features of what became in 1966 the London Transportation Study. However, the first of the newly sponsored, fully fledged transportation studies was the West Midland Study, where data collection began in September 1964. The approach spread through the conurbations (including Greater Manchester, Merseyside and Glasgow, where the more simple origin and destination type of traffic survey had already taken place), and then spread to the smaller cities and towns. By the early part of 1966, eighteen English studies with costs shared by the Ministry were under way, the smallest of which was in the Somerset town of Weston-super-Mare.[13] *

In the larger centres of population, studies continued to start at various dates throughout the 1960s (and, in one or two cases, in the 1970s also) and, as a result, there was a considerable degree of evolution in the approach and techniques used. In fact it would be inappropriate to draw too fine a distinction between the earlier conurbation traffic surveys and the new studies, as a number of features were common to both. Moreover, one striking, and, as it turned out, significant characteristic of the conurbation studies was the length of time each one took. As a consequence, a number of such studies (London in particular) changed character as they progressed, reflecting to a degree the changing circumstances of the 1960s. Nevertheless, the *basic* structure of the major land-use/transportation studies remained unchanged throughout. This was, in simple terms, to examine the relationship between urban activities and travel, and then, by anticipating how the former were likely to change, to assess also the consequential changes in the patterns of travel and traffic. This would then allow for an assessment of the scale and location of required investment in new highway facilities.

At the core of the analysis was the household, which American experience indicated was either the start or finish of nearly all person-journeys. Consequently, the so-called home-interview survey was at the heart of the data-collection process and was one feature in particular that set the transportation study apart from the earlier, more primitive, traffic surveys. It also brought a new dimension to the costs of traffic planning. In London, for example, nearly 50,000 households were interviewed, but even in comparatively small cities like Belfast some 16,000 homes were approached for information. The information sought included the number of daily journeys by persons living in the household, and their social and economic characteristics, such as the size of family, the number of household members employed, incomes and ownership of vehicles. Collecting data of this kind for London alone cost almost £500,000, or as much as the LCC had spent on road improvement in the mid 1950s. As one commentor remarked,[15] to suggest spending on a survey on this scale was itself revolutionary.

*In spite of such examples as Weston-super-Mare, the transportation study remained for much of the 1960s a particular feature of the conurbations and large cities. It was not, in fact, until after the Ministry of Transport's 1968 request for local authorities with populations over 50,000 to submit traffic and transport plans that the smaller authorities began to play an active role in the process. Although the thrust of the plans called for in the circular was essentially short-term, many authorities appear to have been spurred on to set these in the context of longer term land-use/transportation planning. One serious impediment to the preparation of these plans was, as Lady Sharp discovered in her manpower survey,[14] a lack of appropriately trained local authority staff.

Subsequent analysis of this data suggested that car ownership and income were of crucial importance. As household income rose, the number of daily trips the household made rose also in homes both with and without cars. But, in addition, and more importantly, at any income level the trip-making propensity of car-owning households was much higher than in households without cars. In London in 1962, it was about twice as high and, as would be expected, virtually all of these extra trips were made by car. This was the crux of the matter. By anticipating the number and location of households, their incomes and especially their ownership of cars for ten, fifteen or twenty years ahead, one could forecast the bulk of future urban trips. (A basic assumption made of course was that future households would act like present-day households with the same characteristics.) It was then a question of making fine adjustments for factors such as multi-car-owning families with their marginally lower car-utilization rates, and for household employment structures.

Of course, in the middle years of the 1960s, when this type of analysis was being conducted in most large cities, there was still a definite optimism regarding prospects for economic growth. Conurbation studies, for example, based their income forecasts on increases of around 3·5 per cent per annum, whilst anticipated car-ownership growth rates of at least twice this rate were common. The consequences of these assumptions for the initial forecasts of car use were nothing short of dramatic. In the West Midlands conurbation, for example, journey-to-work car traffic was expected to increase by almost two-thirds, and other car traffic virtually to double over a fifteen-year period. In Clydeside, over a quarter of a century, journey-to-work car traffic was forecast to treble, and other car traffic to more then quadruple. In London, the 4·1 million car trips of 1962 were expected to increase to more than 9·0 million in 1981.

The implication of these figures as they emerged from the early stages of the computer analysis were alarming. In the conurbations especially, they implied very high levels of road investment indeed and clearly were unrealistic, in spite of the growth ethos of the time. The general picture in the later 1960s, therefore, is one where the analyst in the conurbation studies attempted to control the situation by trying to reduce the forecasted traffic to fit a more limited highway capacity. But, given the limitations of the methods used by the transport planner, this was no easy task.

In one of the first British critiques of transportation studies, published as early as 1964, the Economist Intelligence Unit (EIU) had drawn attention to "certain shortcomings which demand serious attention"[16].* The basic point made by the Unit was that the forecasting methods then being adopted, implicitly assumed that the costs and ease of moving around cities in the future would be no different from conditions at the time that data were collected. But in the

*Other early critiques included Beesley, Blackburn and Foster[17] (Beesley and Foster were to become Chief Economic Adviser and Director General of Economic Planning respectively at the Ministry of Transport) and Plowden.[18]

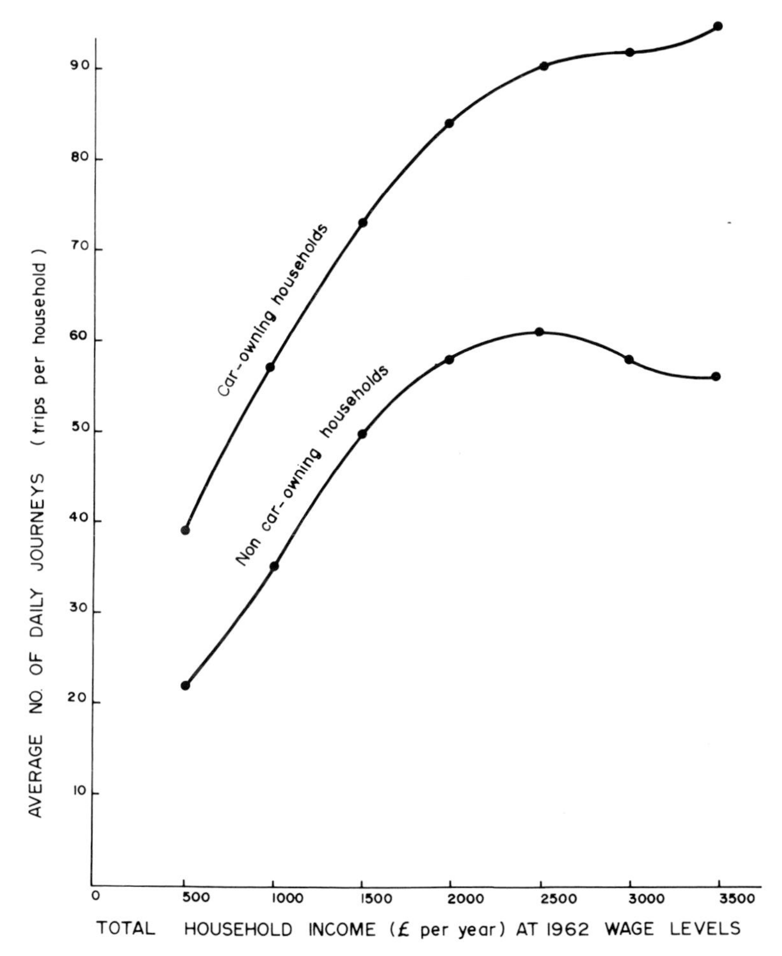

FIG. 7.2. Impact of car ownership and higher incomes on daily travel: the relationships shown are those for London in 1962. It was then thought that rapid growth in incomes and car ownership would have an explosive impact on future urban travel.

absence of very substantial road investment (and in many cases even then) future travel conditions would be much worse, and, therefore, the amount of travel very different from that forecasted. The EIU in their critique added a prophetic comment that it would be unrealistic to prepare forecasts assuming no change in travel costs and, face to face with the problem of excess demand, to cut back arbitrarily the level of traffic. In the event, the methods adopted to reduce the initial traffic forecasts varied from study to study, none were convincing, and some, as the EIU had prophesied, were indeed quite arbitrary.[19]

In London, for example, journeys restrained as a result of inadequate capacity

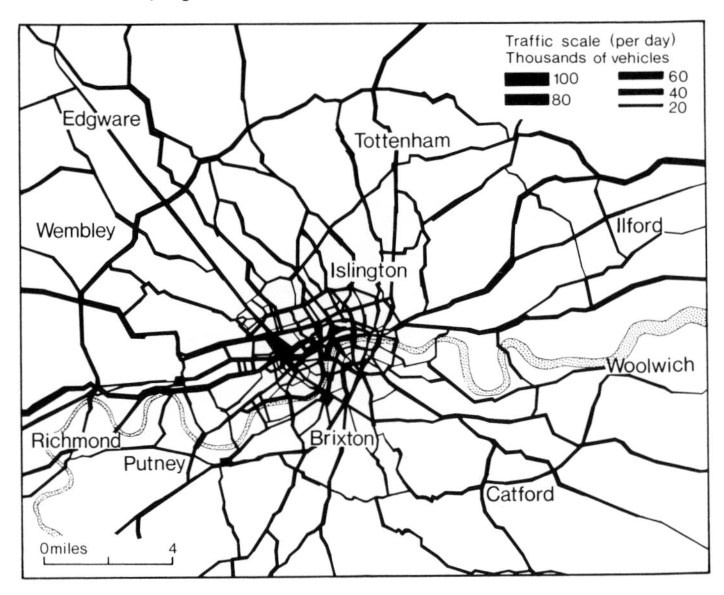

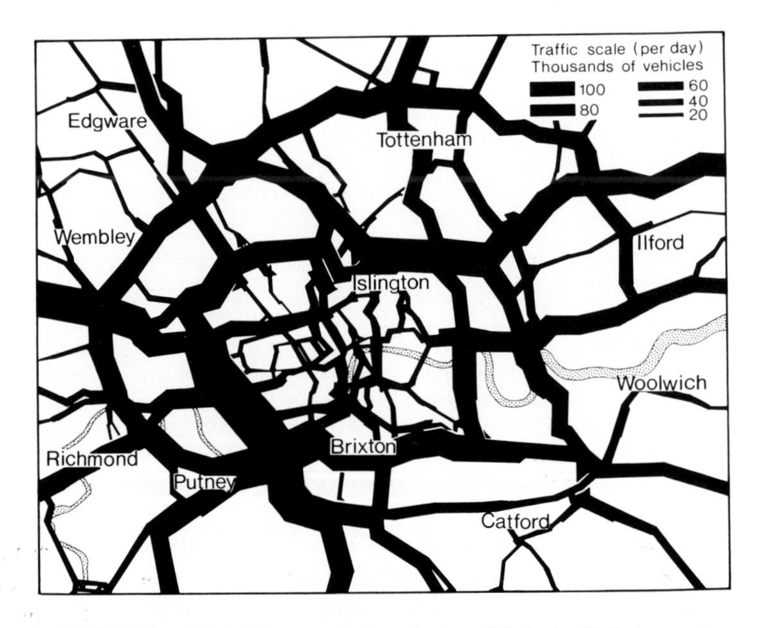

FIGS. 7.3 and 7.4. The expected explosion of daily traffic between 1962 (Fig. 7.3) and 1981 (Fig. 7.4) in London: the data shown are based on analyses in the early sixties by London's traffic planners. It was typical of what was expected to happen in most large cities over a similar period.

of the road system, were divided into two groups: those made by space-saving buses (or by rail) and those not made at all. But the means by which the numbers in the two groups were arrived at owed little to evidence of how people would actually respond to highly congested conditions.[20] As a consequence of these recognized weaknesses, considerable effort was made – in the Greater Manchester (SELNEC) study, for example, where the Minstry's own experts were heavily engaged – to develop methods which would show how traffic "overspill" could be diverted to improved public transport.

These efforts were greatly influenced by the University research of the present Manager of London Transport's bus fleet, David Quarmby, who after leaving university joined the Ministry group associated with the SELNEC study.[21] Quarmby's approach consisted of determining, for those people who had a car available for use, how they would divide themselves between using their cars and using public transport; people without access to cars were regarded as "captive" users of buses and trains. For those more fortunate, less restricted travellers, the basic postulate was that they would choose according to the relative merits of each mode. In practice, this meant that the performance of competing modes was expressed as a "generalized cost". By knowing the value placed upon travel time, time taken can be expressed as so many pence and added to expenditures on fares, petrol, parking-meter charges, in order to work out which mode is cheaper over-all. In addition, the mathematical probability for an individual to choose one mode in preference to the other for a given difference in generalized cost was calculated and incorporated into the analysis.

Unfortunately, there was a basic incompatibility between the detail required for Quarmby's type of approach (for example, realistically representing walking times to bus stops) and the strategic nature of the transportation study with its large-scale view of the city. Either the necessary detail was distorted by meaningless averages, or the study became still more complex and time consuming. Consequently, there were doubts as to the efficacy of this work, too.[22,23]

Endorsement of Earlier Plans

In spite of these attempts to introduce a degree of traffic restraint into the transportation study analyses, when the road plans began to emerge from the first batch of studies in the late 1960s, it was evident that the scale and size of the proposals were large by any standards. In terms of the admittedly inflated prices of the present age, billion-pound schemes were common currency in most large cities. The most telling characteristic of the plans, however, was not that they were large but that they were also comparable in scale and in detail to the blueprints that had emerged from the development plan review process earlier in the 1960s.

Total road expenditure proposed in 1969 in Liverpool, for example, was, after allowing for different price bases and time scales for construction, broadly of the same order of magnitude as the figure in the 1965 plan.[24] In Belfast the road system proposed for 1986 in the final report of the study published in 1969 differed very little from the main schemes under consideration by highway authorities in 1967, except that one of the proposals, for a bifurcation of the south-eastern motorway, was deleted. In the Greater Manchester area, almost 80 per cent of expenditure proposed in the SELNEC study was upon schemes in preparation prior to the release of the study report. In London, the affinity of the preferred (Plan 3) network, of three ring motorways linking a number of new radials, with the Abercrombie Road Plan of 1944 was evident. But an equally close parallel existed with the GLC road plan that had been published in 1967 in a booklet entitled "London Roads – a Programme for Action". A later GLC publication was to add that "From the broad results of [the Transportation Study] it was deduced that the size and scale of the primary road network proposed by the GLC were not fundamentally wrong".[25] Similar sentiments were expressed north of the border where Glasgow's Highways Convenor argued in 1969 that "The city's long term Highway Plan . . . has been fully vindicated by the Greater Glasgow Transportation Study".[26]

This was not, of course, what Buchanan had intended when promoting the land-use/transportation study as the basis for planning a motorized yet civilized urban environment. Buchanan's approach to the traffic problem was first to delineate the areas in which life was led and activities conducted, and, as this cellular structure took shape, the pattern of highways, falling in the troughs between the environmental areas, declared itself. A good start with this approach had been made in the Leicester Traffic Plan. The Leicester study made extensive use of American transportation study techniques and yet appeared to subjugate the computer print-out by first identifying eight areas of distinct character and with distinct boundaries. *The Economist* commented, "soon, one hopes, this approach will be a routine exercise for all cities".[27] But these hopes were dashed. In the application of the strategic transportation study methods imported from North America, the necessary sensitivity of application was too often missing. Somehow Buchanan's philosophy was turned inside out. In too many cases it appeared that the transportation studies were used as vehicles merely to confirm, endorse or demonstrate that what was already on the drawing board or in the pipeline was indeed broadly correct.

Conclusions

With the advantage of hindsight, it is now apparent that the transportation studies of the 1960s incorporated wildly incorrect, highly optimistic assumptions on such fundamental matters as population change, the growth of disposable incomes and car ownership. An analysis conducted at the end of the seventies suggested that the forecast error over ten years in thirty-one early transportation studies was, on average, around 12 per cent for population and

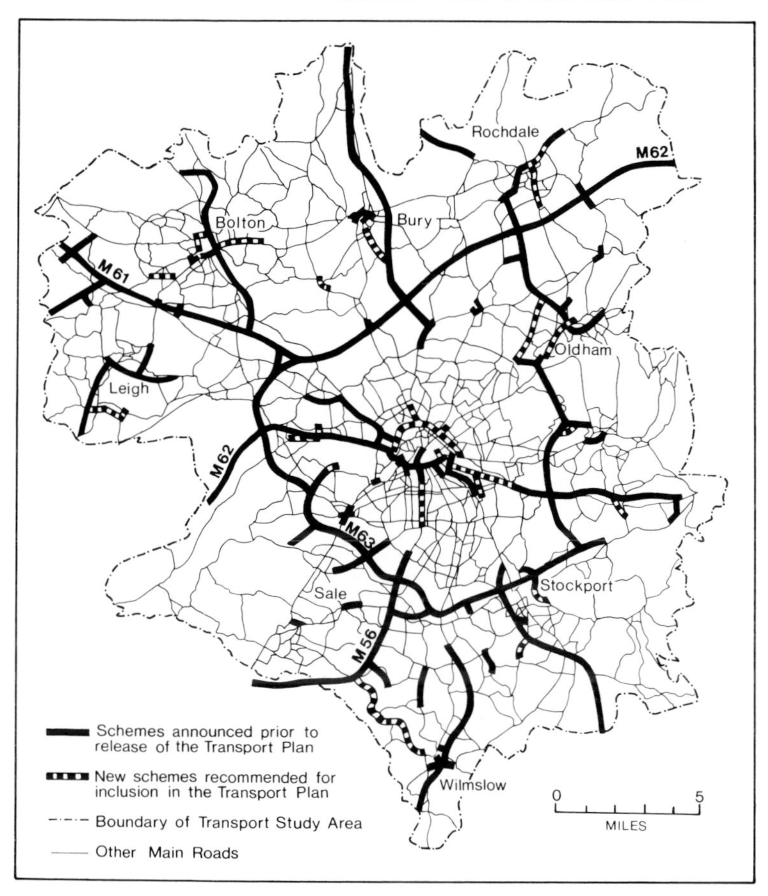

FIG. 7.5. The dominance of committed road schemes in transportation study plans: this example comes from the South-East Lancashire–North-East Cheshire study centred on Manchester, but is typical of the proposals in other conurbation studies in the late 1960s.

employment, and over 20 per cent for cars per head and household income.[28] In fact the startling conclusion arrived at was that if the planner had assumed "no change", frequently he would have been less inaccurate. Of course, it has to be remembered that, at the time, an overwhelming emphasis was placed by both politicians and bureaucrats on faster economic growth, and in many instances on a somewhat naïve belief that establishing appropriate physical and economic planning mechanisms would alone bring this about. In addition, until the mid sixties, Britain's population was growing rapidly, and the subsequent decline in the birth rate was one of those events of nature that could not be foretold with certainty.

There were, nevertheless, early warnings of the demographic and economic down-cycles. In 1968 for example, John Blake, a research officer with one of the London Boroughs, had placed a question mark over many of the London Transportation Study assumptions in an article "The Growth of Traffic in Greater London: Buchanan and London Traffic Survey Too Alarmist?".[29] He pointed amongst other things to the fact that the population of Greater London had fallen and not risen as anticipated, and that the proportion of car-owning households was rising at little more than half the rate expected. The previous year, attention had been drawn to the disparity, also in London, between employment trends and predictions.[30]

But one wonders, had it been possible to foretell with greater accuracy future population, employment, income and car-ownership growth, whether the transportation study proposals would in any case have been much different? The scale of the proposals might have been less and the proposed construction period longer, yet the fundamentals of the motorway network would surely have remained unaltered. The die had been cast already. As one witness told a House of Commons Select Committee in 1972, at the time the studies were conceived "there was an expectation amongst the local authorities and politicians, and to some extent amongst the public, that certain highway schemes would be built".[31] Many of these schemes had been around a long time, and very often they were regarded as "committed". And once these committed schemes were placed on the map, those that were not-quite-so-committed nevertheless fell into place too. In these circumstances, there is perhaps more than a grain of truth in the wry comment of a former Glasgow City Councillor that transportation studies provided powerful underpinning for the less rigorous recommendations of planners and engineers. For the mystified councillor, they gave weighty evidence of "scientific" proof which helped to banish intuitive doubts.[32]

CHAPTER 8

Urban Roads: The Environmental Backlash

BY the end of the sixties, the shift of emphasis in the roads programme that the Government had long intended was in sight. The 1000-mile programme of inter-urban motorways was all but complete, and, in the cities, reports of the land-use/transportation studies were emerging in significant numbers and, as the executive put it, demonstrating the need for continuing major investment in urban roads.[1] The stage was now set for diverting an increasing proportion of a still rapidly rising roads budget to the towns and cities and launching a major programme of urban road construction. Unfortunately for the ambitious road planner, there was one ominous and telling sign: the tide of public opinion appeared to be turning against such a large-scale programme.

The Visionary Preface

Until the mid 1960s urban motorways as a concept appeared to enjoy general acceptance and a still surprisingly wide measure of support. The road lobbies for many years had, of course, canvassed their case. In 1956, for example, the British Roads Federation had organized in London a large international conference on the subject.[2] Dubbed at the time a "Highway United Nations Organization", it had the express purpose of putting pressure on Harold Watkinson, then Minister, to get on and construct urban motorways, a resolution having been passed to this effect at the Conference. The lobbies also organized public competitions such as the Roads Campaign Council's 1959 challenge "New Ways for London". The winning proposal, submitted by Alan Proudlove, later to become transport professor at Liverpool University, was based on a system of motorways tangential to the centre of London.

But, more significantly, the media joined in with proposals of their own. In 1956 the *London Evening News* launched with customary ballyhoo a scheme for a riverside highway running from Battersea Bridge to the Tower (and appropriately sunk in a cut-and-cover arrangement in front of Parliament).[3] Harold Watkinson described it as a "bold and far sighted idea".[4] In the early sixties it was the *Observer's* turn, with its transport correspondent putting forward a 170-mile hexagonal motorway network for the Capital.[5]

This mood of confident optimism continued well into the sixties and was to be reflected, too, in broad based support for urban motorways at the political level. The Conservative Bow Group argued in 1960 that new high-speed roads within urban areas were essential if the traffic problem was to be resolved, but, they added, roads needed to be sensitively sited, and related to more general problems of movement.[6] Such moderation was not to be found, however, in the 1963 report of a transport study group meeting under the auspices of *Socialist Commentary*.[7] In what was to prove a remarkable testimony to a view widely shared at the time when the first transportation studies were being conceived, the Labour group considered that "the long-term answer to the traffic problem [was] to reconstruct the cities" and that "cities should be designed within a broad framework of urban motorways". Their only caveat was that it would be erroneous to start large-scale renewal before the results of more basic research were available. However, as events were to show, by the time the transportation studies had produced these results (and the Government had money available to start implementing them) the climate of opinion had changed dramatically.

Pressures

There had been, even during the optimistic years up to the mid sixties, portents of what was to come. There were, for example, the celebrated controversies of the Bath Tunnel and the Meadows Road in Oxford. In the case of Bath, Colin Buchanan's private consultancy practice had been commissioned in 1963 to carry out a planning and transport study. The key to Buchanan's resulting plan was a twin two-lanes-each-way tunnel under the very heart of the heritage area of central Bath. But for many, this proved to be too bold and too expensive a solution, and in 1966 a furious debate ensued with the Council unable to reach a decision.[8]

The city fathers of Oxford were less equivocal than those of Bath, but nonetheless faced strong opposition to their plans to relieve the central part of Oxford from traffic (thereby restoring and preserving the unique character of Oxford's historic centre) by building an inner relief road. For part of its route the planned road crossed Christ Church Meadow, an area of parkland between the colleges and the confluence of the Thames and Cherwell, and it was this aspect which was strongly opposed by many influential bodies, not least of which was the University.[9] After a special enquiry in 1960 firm proposals for the road were submitted as an amendment to the City Development Plan. This was subject to a normal public enquiry in 1965 and subsequently referred for further study by Richard Crossman, Minister for Local Government. But then Oxford, like Bath, was a special case.

These mid years of the sixties were probably the time when in more mundane places than Oxford and Bath, attitudes to new roads began to change decisively.

The rapidly rising roads budget was now allowing for more urban spending (in spite of the focus on the rural motorway programme), and in more than a dozen such cities, people were beginning to see what large-scale urban road plans implied, by observing road construction at first hand. During 1964/65 alone, thirty-two schemes on classified roads were completed in London and the county boroughs, and 101 more were in progress.[10] Consequently, within such diverse centres as Chester, Burnley and Manchester, large elevated road structures were emerging. But in these stakes of advanced engineering, the prize was taken by a section of the M4 snaking through West London and incorporating the longest road viaduct in Europe.

With perfect timing, the day following the opening of the M4 viaduct, *New Society* revealed (in a story syndicated to the front page of the *London Evening Standard*)[11,12] what was to prove a critical element in London's road plans, namely a box-shaped ring motorway elongated in an east–west direction and running about four miles from the edge of the central area. (Peter Hall writing in *New Society* had in fact scooped a report to be submitted to the GLC by the Planning and Highways Committees on its day of investiture, April Fool's Day 1965, seeking to safeguard the route of what, henceforth, was known as the "motorway box"). Now, for probably the first time, the image of a motorway skirting the roof tops of Chiswick, and the image of the *Evening Standard*'s map portraying fifty or so more miles, began to bring home to hundreds of thousands of people exactly what coming to terms with the motor-car might mean in reality.

Henceforth the mood certainly began to change. The generally neutral tones with which the *Standard* and other papers had used to report on the "motorway box" in early 1965, soon began to give way to more adverse scripts, as opposition groups formed and wrote letters to newspaper editors.* Headlines such as "And London may lead the way"[15] gave way to "Group fights 'Premature London Motorway Plan'"[16] (1966), "A reprieve from motorway threat"[17] (1967) and "The rape of Old Cheyne Walk" (1967).[18] By the time the motorway box had been incorporated into the official Greater London Development Plan in 1969 (and relabelled Ringway One), an anti-motorway crusade was in full swing and most of the press had come down on the side of the crusaders. The press, with a keen instinct for popular anxieties, now devoted whole-page stories to the issue, and in some cases with headings such as "How Safe is Your Home?",[19] played

*Good general reviews of the London motorway saga are to be found in Douglas Hart[13] and Peter Hall.[14]

FIG. 8.1. An aerial view of central Oxford showing the proposed line of the controversial relief road crossing Christ Church Meadow (centre) and joining up with a completed section of the inner ring road (left). The long curve of the High Street and Magdalen Bridge can be seen clearly (top right) leading to the residential suburb of Headington (right). The confluence of the Thames and the Cherwell lies at the bottom centre (*see over*).

FIG. 8.1.

FIG. 8.1.

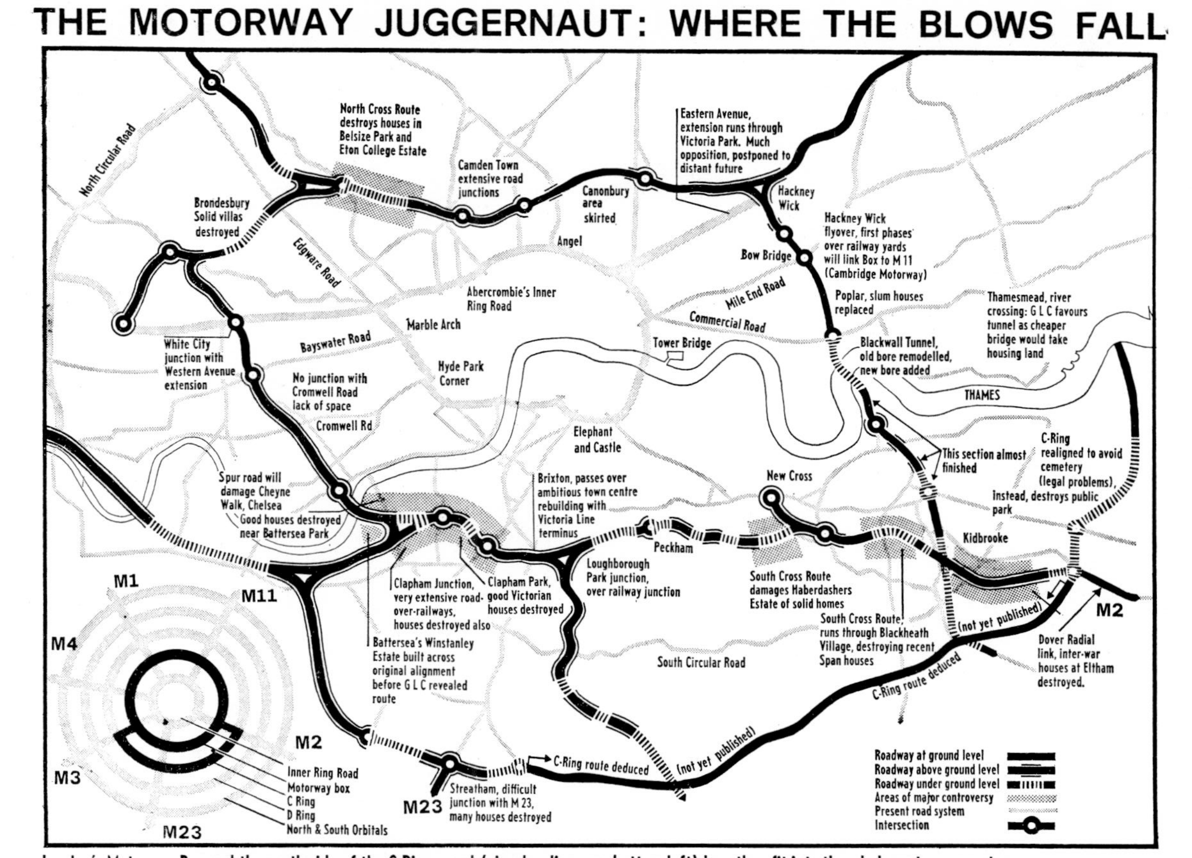

FIG. 8.2. "The Motorway Juggernaut: Where the Blows Fall": sections of the press played on community anxieties concerning motorway schemes. This is one example from the late sixties.

on the point made by *The Times*' Transport Correspondent that the motorways had become "a focus for fear".[20]

Such fears were soon aggravated when, with timing disastrous for the road advocates, one of a series of radial motorways feeding into the ringway system opened on the eve of the enquiry into the Greater London Development Plan, which had as its backbone the same ring of motorways. Although the physical and engineering circumstances of "Westway", as the controversial road was known, were little different from those that prevailed when the Chiswick to Langley section of the M4 opened five miles to the south west, five years earlier, the political environment was now very different. The row that followed the opening of Westway when the residents of Acklam Road were subject to the full blast of the traffic nuisance only feet away from their bedroom windows, again focused public attention by TV camera, microphone and press photographs on what such a road would cost in social terms. Five years on, it was a public more acutely conscious and sensitive to such issues. As one commentator subsequently argued, Westway probably did more than seven years' reflection on *Traffic in Towns* to ensure that politicians would not in future be allowed so flagrantly to ignore the environmental impact of road schemes.[21]

In the Welsh capital, Cardiff, a similar dispute was being played out to similar rules.[22] Here the local land-use/transportation study reported in April 1968 with recommendations which included twenty-four miles of primary distributor road to be built to urban-motorway standards. At the focal point of the proposed network was "the Hook road, a 100-foot wide, six-mile long swathe of concrete and flyovers and flyunders" requiring the sites of 1500 homes and acting as a major spine linking the central area (there was a strong link between the proposed road network and city-centre development) to major east–west roads lying to the north.

The scheme went to full council in the autumn of 1968 and was carried by a substantial majority and became official policy. At this point a powerful opposition organized itself under the spearhead of the Cardiff United Residents Association (with an estimated 6000 members) and, as in London, invoked the emotive issue that "homes should always come before cars and concrete". There were, however, two important contrasts with the London situation: in Cardiff the local press was pro-road and the issue entered the realm of party politics at a much earlier stage when the Plaid Cymru, Co-op. Party and Cardiff Socialists came out against the Hook.

In several English provincial cities, events were taking shape along similar lines, but rather more slowly.[23] In Portsmouth an extensive highway plan for the city had been proposed in 1965 and approved in principle by the full Council in October 1966. By the end of the following year, detailed designs of a ten-lane north–south motorway and western sections of an east–west route between Portsmouth and Southsea were well advanced. About the same time opposition began to form to the north–south route which involved the demolition of between 200 and 300 houses. The route was fought by a residents association which narrowly failed to win the ward seat in local elections.

In neighbouring Southampton, the focus of controversy was the Portswood link, connecting the M27 with the dock approach road to the west of the city centre. About 650 homes were required to complete this plan, and in 1967 a community action group was formed. Nottingham's highway plan, published in 1966, included nearly twenty-three miles of improved existing road and eleven miles of new road. Most important in the latter category were the Eastern Bypass and Sheriffs Way, around which had been designed a redevelopment scheme for the St. Ann's area of "slums". Here the principal objections came from a tenants' association which had been formed to fight comprehensive redevelopment prior to the release of the 1966 highways plan. Generally, however, in these provincial cities throughout the sixties, opposition to the road plans was local in character and, perhaps as a result, focused on narrower objectives than opposition in the capital cities. Nevertheless, here too the warning signs were evident.

Palliatives

In some respects the Government appeared to be very quick in reacting to

"While everybody else was complaining, I decided to make the most of it."

FIG. 8.3. A comment by *Punch* on urban motorways.

these mounting pressures (in other cases they moved much more slowly). One speedy reaction was to change the long-held commitment to diverting, in a relative sense, resources from trunk roads to urban roads.

In 1969 the Government had affirmed that one objective of the roads programme for the first few years of the 1970s was "to devote a substantial and growing part of the programme to road works in urban areas".[24] This was reaffirmed in the May 1970 White Paper, "Roads for the Future". But faced with the growing prospect of resistance to urban road building, the policy was quickly revised.

The opportunity to make the revision came with the review of public expenditure following the election victory of Edward Heath. The Conservatives had been elected on a manifesto of cutting public expenditure and now they sought to put this into practice.

On the whole, roads did not fare too badly in what was purported to be a substantial reduction of previous plans for public spending. The maximum cut-back for roads of about £25 million was planned for 1974/75 and represented about 2 per cent off the previous programme. Of this, about £20 million was to be cut from capital expenditure on major roads and most of this was to fall in the largely urban, principal roads category.

This change of plan was confirmed in the following public expenditure White Paper of November 1971[25] which set out the proposals to increase expenditure on road construction and improvement by 6·2 per cent a year compound (in constant terms) from 1970/71 until 1975/76. But for English trunk roads and motorways, this same construction and improvement component was set to grow on average at a faster rate of 10·5 per cent.

There was no evidence that resources were being redirected towards trunk roads, and, importantly, urban roads still remained a growth sector. Nevertheless, the cuts now taking place, small though they were, did mean a marked shift in emphasis. The long established policy, first mentioned in the mid sixties, of boosting spending on urban roads relative to trunk roads, had, in effect, been quietly abandoned.

Another major reaction of the Government to the evolving environmental backlash was to take steps to make urban motorway building more effectively part of a wider process of planning. Previously, designing and building new roads had been treated, in spite of the frequent links with comprehensive redevelopment, largely as a self-sufficient engineering exercise. Now steps were taken to change this, and at the same time to change fundamentally the rules determining the amount of compensation that could be paid to persons affected by road schemes.

At the time that Westway brought home to many the horrendous impact that urban motorways could have on those living alongside, highway authorities were severely restricted in their ability to compensate such losers. Moreover, they were restricted also to buying property essential to the building of the actual road. They could not, for example, buy affected adjacent property nor buy excess property in order to establish a *cordon sanitaire*. To circumvent these limitations in connection with their proposed inner-ring motorway, Liverpool Corporation promoted a Private Bill in the mid sixties, but others,[26] including the GLC, had been pressing the Government for a general legislative change.

In June 1969 a large committee was established to consider the issue. Its initial chairman was Sir James Jones, later to become Permanent Secretary at the Department of the Environment. In addition to officers from the Ministries of Transport and Housing and Local Government, the committee included a long list of distinguished external members drawn from the ranks of academia, local government and the architectural, town planning and engineering professions. Supporting the committee was a team of officials known as the Urban Motorway Project Team, set up on a full-time basis to do all the technical work and to co-ordinate the work of consultants. The latter were engaged on a large number of technical case studies. The terms of reference of the main committee were first to examine existing policies used in fitting major roads into the urban fabric, second to consider changes that would improve the integration of roads and their surroundings, and third to examine the consequences of such changes for public policy, statutory powers and administrative procedures.

Backed by extensive research, the recommendations emerged finally in the July 1972 Report of the Urban Motorways Committee, *New Roads in Towns.*[27] Amongst the many recommendations of the Committee were suggestions for designing roads and surrounding areas as one operation, and then including the costs of remedial measures as part of the cost of building the road. To enable this to be done, the Committee suggested changes in powers of highway authorities to enable acquisition of land adjacent to the new road and to enable expenditure on better road design and on the sound-insulation of buildings.

The Report, together with a full-scale review of the Compensation Code, formed the basis of a White Paper, *Development and Compensation,*[28] presented to Parliament in October 1972. The resulting legislation, the Land Compensation Act 1973, introduced a number of far-reaching reforms of the Compensation Code.

The provisions of the Act allowed for much-extended powers and for expenditure by authorities to mitigate nuisances from *new* motorways and roads. For example, authorities could now acquire land and property to enable road schemes to include noise barriers, landscaping and earth embankments. Where such mitigating works were not feasible or were not fully effective, and where additional carriageways increased traffic noise (by one decibel or more, above a minimum of sixty-eight decibels, for 10 per cent of the period of time between 6 a.m. and midnight on any normal weekday), authorities were *required* to sound-insulate residences.*

*The Act does not apply to Northern Ireland: it was extended to Scotland by the Land

If measures of this nature proved inadequate, the Act provided a statutory right to compensation for nuisance from the *use* of new public works, thus modifying a rule followed for over a century. The compensation was based upon assessed, significant falls in the value of property caused by concomitant noise, fumes, smoke, vibration and artificial lighting. In addition, there were changes in the assessment of compensation for compulsory purchase, and the provision of a Home (and Farm) Loss Payment as a mark of recognition of the special hardship created by compulsory purchase procedures. In brief, the new policy required transport developments to be planned in such a way as to minimize the disturbance and disruption they caused, and for any distress which remained to be alleviated by improvements in the Compensation Code.

But such fundamental changes did not come cheaply. Based on some 100 miles of new motorway and a further 150 miles of other major new roads or substantially improved roads opened each year, one estimate[30] put the total annual cost of improving road designs and implementing the land acquisition, sound-insulation and compensation provisions of the 1973 legislation at £60 million–£70 million (1972/3 prices). Thus, as spending on constructing and improving urban roads (where most of these costs would fall) was, at the time, around £200 million,* the new provisions were of consequence not only in terms of revising amenity rights, but also in terms of road expenditures; henceforth building new roads in towns promised to be a significantly more expensive proposition.

Conclusions

During the sixties, the change of public mood with the regard to urban road building was profound, and, in spite of the focus of the media on reactions to London's ringways, it was a change of mood by no means exclusive to the Capital. The cause of this change was difficult to determine. Undoubtedly, a contributing factor was an increasing number of new urban roads (completed or under construction) to judge and to react to (if not at first hand then through the newspapers and television). And sometimes what the public saw – Westway abutting Acklam Road being a classic case – were the more brutal examples. Increasingly, therefore, people were able to interpret the lines on the map into a presumption of reality, and often the reality appeared very different from the vision of the late fifties and early sixties.

Compensation (Scotland) Act 1973. The provisions of the Act do not relate solely to roads or indeed to transport developments alone. All new public works used in the exercise of statutory powers were included, although highways and aerodromes were expected to constitute the real villains of the piece. The details presented here are in accordance with the Noise Insulation Regulations 1975.[29] Also provided were *discretionary* powers to insulate against noise from road construction work and from traffic on *altered* roads.

*Based on figures for urban road spending given in House of Commons, Report on "Urban Transport Planning".[31]

It would be wrong, however, to think of the changed mood and the cause of it in the context of urban roads alone. One could detect, for example, a growing concern with environmental issues in the late fifties when, significantly, two groups later deeply involved in the transport debate were both founded; the Civic Trust in 1957 and the Victorian Society in 1958. But at this time the major planning controversies involving amenity appeared to focus upon violation of the countryside, often by the power and water authorities, and, perhaps surprisingly, only occasionally by the motorway planners.

Eventually this focus of environmental controversy shifted towards the towns. When it did so, it reflected the growing momentum behind the post-war property boom and the processes of comprehensive redevelopment. New roads were usually part and parcel of these redevelopment schemes; distributors routed through the soft areas of slum housing or inner ring roads closely tied to a new commercial heartland. And when the transportation studies had completed their work, many of these road plans were upgraded into motorways and given added emphasis in the overall plan. Nevertheless, the reaction was rather more general than simply a revolt against urban motorways; it was a reaction also to new shopping precincts of glass, concrete and plastic and the sweeping away of neighbourhoods to be replaced by industrialized blocks of flats often twenty or thirty storeys high.

In the late sixties, urban policies began to change to accommodate this new mood. For example, the 1967 Civic Amenities Act (promoted as a Private Member's Bill by Duncan Sandys, President of the Civic Trust, and passed with Government backing) gave statutory recognition to the concept of conservation areas. Now whole areas of towns, as distinct from individual buildings, could be protected if the buildings *en masse* provided a character worth preserving. Housing policies, too, were moving emphatically towards modernization and rehabilitation.

There was, however, little evidence at the beginning of the seventies that the full significance of these changes of public attitude had caught up with the Ministry road planners. For years they (and the transportation studies promoted by them) had proceeded on the basis that comprehensive redevelopment was inevitable. It appeared they continued to do so. Their basic response to the anti-road campaigns was to think in terms of "sugaring the pill" – to place faith in the possibility that opposition to road schemes might evaporate with better road design and more generous compensation. What was not evident at this stage was a change in the basic belief that the problem of traffic in towns in the long-term could be solved only with the aid of a substantial programme of building new roads.

CHAPTER 9

Town Traffic and Roads: Winds of Change

WHILST the Urban Motorways Committee deliberated for nearly three years, events continued to unfold. The prospect of more thoughtfully designed roads and more compensation did not appear to blunt criticism of urban motorways, and during these early years of the seventies, opposition to urban road plans continued to mount. However, as time passed the opposition began to take a different form.

In the 1960s, opposition to the road proposals had been largely negative. The reaction was a gut reaction and quite often the injured party was interested only in moving a scheme into someone else's backyard. There were few constructive suggestions for alternative solutions. But this situation too began to change and soon there emerged an alternative philosophy to the problems of city traffic.

An Alternative View

An early and cogent expression of this was to be found in a Report titled *Motorways in London,*[1] published in 1969. At the invitation of the London Amenity and Transport Association (LATA), the main group of lobbyists against the ringway plans, Michael Thomson was invited to head a multidisciplinary working party. The thrust of the working party's argument was that the GLC had seen the problem too much in terms of keeping traffic moving and too little in terms of the quality of movement or, for that matter, the quality of life. In the eyes of the working party, the problem could not be attributed to a past failure to build more road space, but more widely to a failure and inability to apply economic criteria to transport. Their conclusion was that the basic need in London was "to produce a selective combination of improvements in facilities on the one hand, and shifts in demand on the other, with the aim of bringing the whole transport system gradually into balance".

Both a focus and a stage for this viewpoint of the working group was provided the following year, 1970, with the start of the inquiry into the Greater London Development Plan. The preparation of this plan was one of the chief statutory responsibilities of the GLC, and after several years of preparation it was submitted to the government for approval in 1969. The inquiry, set up to consider the many objections to the plan, was conducted by a panel headed by Sir Frank Layfield, a distinguished specialist in planning law. The GLC's proposals for the ringway system of orbital motorways with their supporting radials formed the backbone of the plan. Consequently, about three-quarters of the objections (which exceeded the astonishing number of 28,000) constituted disagreements with the transport proposals, and especially with the road proposals.

The inquiry was to last for more than two years (as one commentator noted, it rivalled the Roskill Commission on the third London airport as the most protracted planning inquiry in British history) before the Panel submitted its report to the Government in December 1972.[2] On the issue of the road proposals, the Panel felt that the motorways might generate an unacceptable amount of traffic in Central London unless strict management and stringent restraint were applied. There was no evidence in the Plan that such ameliorating policies were intended. But, if such controls were introduced in parallel with motorway construction, then the provision of orbital capacity was too lavish and would be underutilized. Therefore, the Panel proceeded to pare down the road proposals leaving as major features only the M25 (which for the most part lay outside the GLC's area of responsibility) and the inner, and most controversial, ring. The Panel found the latter to have the highest rate of return, which, although inadequate by usual standards, was thought to be enhanced by important environmental advantages gained by canalizing traffic which would otherwise use the secondary road system.

This, of course, was diametrically opposed to the case the LATA had presented. The LATA were especially critical of Ringway One because of the large amount of property that would have to be purchased to accommodate it, and because it was thought likely that traffic generated by the motorway would swamp secondary roads. When the Layfield Report came out in support of the innermost ringway, naturally there was an outcry. Nevertheless, there was much in the Panel's report to suggest a sympathetic treatment of the LATA's case.

However, the opinions and recommendations of the Panel were, by now, for the most part academic. For what was to prove the most significant decision had been taken already. In June 1972, after a great deal of apparent heart-searching and vacillation, the London Labour Party finally decided to take an anti-motorway stance. At a time when the popularity of the Conservative Government at Westminster had waned, the Labour Party, with this platform, won the day in the 1973 GLC elections. London's motorway plans were abandoned and destined to join Peter Hall's catalogue of *Great Planning Disasters.*[3]

By now, other cities, too, had abandoned, deferred or substantially modified plans for urban motorways. Nottingham was one celebrated case. Here the first scheduled parts of the 1966 highway blueprint (Sheriffs Way and the Eastern Bypass) required an amendment to the Development Plan and an inquiry to this effect had been held in early 1970. This provided a set-back for the road proposals from which they were not to recover. The inquiry inspector felt that the proposed road schemes were too large in scale to be justified. In what, for the time, proved to be a remarkable piece of unconventional thought, he added that what was urgently needed in Nottingham was a much greater emphasis on the need to control, rather than attempt simply to accommodate, town traffic.

The Minister upheld his inspector's decision and the highways plan was thrown back into the melting pot. The Department of the Environment encouraged the city to submit new road plans as soon as it could, so as to retain the funds already allocated to the city's road programme. However, a change of controlling party in the May 1972 local elections led to thoughts of a radical alternative. This was the "zone-and-collar" scheme announced in October 1972, although not fully introduced for some time after.

The essence of the proposed scheme was the use of traffic lights at peak periods to ration the number of cars entering the main radial roads from suburban areas. This was supplemented by a collar (or cordon) of traffic controls on the main radial roads around the edge of the inner city. In both cases priority measures were to be used to allow buses (in addition to cyclists and emergency vehicles) to bypass the rationing points. Thus, vehicles would be ponded back into organized queues to permit buses in particular to flow freely along central city streets. In the inner city areas, car commuters were to be further discouraged by reductions in the number of long-term parking spaces and by an increase in parking charges. Those motorists using cars for the journey-to-work were thereby given every incentive to transfer to a bus service improved not only by the priority measures but also by increased frequencies, new routes and the generous provision of park-and-ride facilities.*

The really significant point about the Nottingham scheme was that it was seen as an *alternative* to large-scale road building. Its proponents shared a belief that urban motorways were by no means an essential solution for urban traffic problems and that it was possible for city transport to continue to operate well, without greatly expanded road capacity. Their emphasis, instead, was upon managing the demand for existing road space rather than trying, by building new roads, to accommodate as much traffic as possible and then leaving traffic restraint and better public transport to mop up the residue. It was this difference in emphasis that was crucial. However, by this time, 1972, what had started out in the previous decade as an unconventional, if not radical, view was rapidly becoming a view with the support of many august bodies, not least of which was the House of Commons Select Committee on Expenditure.

Parliamentary Pressures

At the end of 1971, the Environment and Home Office Sub-committee of the

*A scheme on very similar lines to Nottingham's "zone and collar" had been suggested for Coventry by the City Engineer early in 1968. The Council's traffic committee at that time accepted the basic concept and authorized discussions with manufacturers of traffic control equipment.[4]

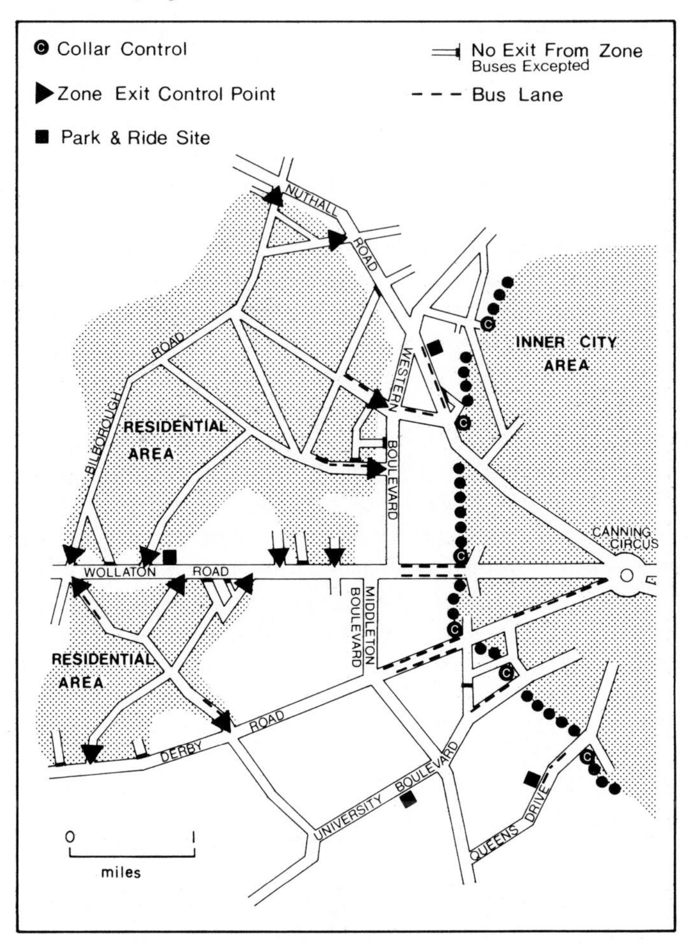

FIG. 9.1. Nottingham's zone-and-collar experiment: the diagram shows the controls operating during the morning peak period.

House of Commons Expenditure Committee decided to undertake a major inquiry into Urban Transport Planning.[5] The decision was made in the knowledge that there was increasing public concern with city transport which was manifest, as the Committee put it, "in increasing complaints about inadequate train and bus services; in action groups and public demonstrations against the building of urban motorways; in dissatisfaction with traffic congestion and the swamping of city streets by private motor cars and intrusive heavy lorries".

The inquiry concentrated chiefly on the journey-to-work situation in the main urban centres. Its main purpose was "to determine to what extent it [the Department of the Environment] is helping by its policies and its financial measures to solve the problems of urban transport planning". The Committee pointed out that executive action in this field rested largely with local authorities, and, as a large part of the finance came from central government, the Department had, they considered, a duty to see that this money was spent wisely.

It was, by the usual standards, a major inquiry lasting for the most part of 1972. Sixty witnesses appeared before the Committee, including the Secretary of State for the Environment and a dozen of his most senior administrators, together with academics, consultants, local authority representatives, members of three conurbation Passenger Transport Authorities and representatives from the police and nationalized transport industries. In addition, there were numerous memoranda of evidence submitted by bodies as various as the British Cycling Bureau, the Committee for Environmental Conservation, the Royal Town Planning Institute and the Association of Municipal Corporations, as well as a number of organizations associated with the roads lobby.

Based on this "impressive array of evidence", as the Secretary of State for the Environment was later to refer to it, the Committee fashioned a fifty-two-page Report that was to prove of considerable significance. Throughout the Report there was a consistency of view and purpose which was well summarized in paragraphs 26–28. Here the Committee reported that they had come to two main conclusions:

> "Firstly, there should be a major and substantial effort to improve public transport . . . Secondly, the use of private cars for the journey to work should be severely discouraged in areas where it impedes public transport."

They went on:

> "We form these conclusions because in city centres the benefits enjoyed by motorists in peak hours are more than offset by the cost and penalties that they impose upon the community as a whole. We do not believe that in the short term an extension of urban road building of itself represents a solution to any of the problems we have discussed. . . . The arguments used to favour road building seem to us to be in error in presuming that the roads which we already have are being used in the most efficient manner . . . we believe that the first step should be to make the best use of the present road system before adding substantially to it."

They then went on to point out that this optimal use of existing roads involved taking into account the costs of congestion and pollution and that to do so:

> "would lead inevitably to an effective reduction in the volume of traffic and to some shift back from private to public transport. The less mobile sections of the population would on the whole find travel easier, and the quality of life in our towns would be improved."

The Committee's specific recommendations, designed to give effect to their basic conclusions, called for a general and pronounced tightening of parking measures geared to the restraint of traffic. (Nine of the dozen specific traffic restraint recommendations concerned parking.) Included was the more radical proposal for encouraging the transfer of private parking space within offices and other commercial buildings to alternative uses, such as storage, by, if necessary, taxing its use by car commuters.

Complementing the traffic restraint measures were proposals designed to promote public transport. Here the focus was upon giving buses priority in the use of road space supported by the more radical proposal of giving rate rebates to firms which staggered hours, thereby spreading the peak load that public transport had to cater for. Less radical, but possibly of greater financial significance, was the Committee's recommendation that Government subsidies should be made available to cover losses on urban bus services but in such a way that managerial efficiency was not undermined.

In its general tone, as well as in its specific recommendation, the Committee's Report provided a powerful indictment of policies geared principally or even moderately just to keep traffic moving. It was an indictment especially of those policies that aimed to do this by exclusively building more roads. Moreover, it was a statement that gave both credence and, at last, a measure of authority to the more articulate and well-considered opposition challenging local plans for urban motorways. In fact, the Report represented and reflected a fundamentally different (but now dominant) philosophy and a fundamentally different world from that in which, less than a decade earlier, it could be asserted with confidence that "beyond question, there will have to be a great deal of urban road building in British cities".[6] *

A Fundamental Shift of Resources

The report caught the mood of the times and was well received both by Westminster and by the public at large. Somewhat unusually, it was the subject of two motions of debate in the House within a few months of its initial presentation to Parliament.[8,9] On each occasion, Members from both sides of the House were generous in their praise, and the Secretary of State for the Environment referred to its "penetrating analysis", although, curiously, Mr. Doig, the Member for Dundee West, an urban area where car ownership was one of the lowest in Britain, found its attitude to cars such that he opposed it.

In the Government's Observations[10] on the report (Observations are the normal way of formally responding to Select Committee reports and, for major

*John Horam, member of the Select Committee and backbencher who was later to become junior transport minister to William Rodgers in the Callaghan administration, personified this popular change of mood. Horam in 1964 as transport correspondent of *The Financial Times*[7] had called for a shift of emphasis in expenditure from trunk to urban roads.

reports, these are usually in the form of a White Paper), the sentiments of the reply concurred in a large number of cases with the Committee's thoughts. The Government defined nine specific recommendations of the Committee as the most important. Of these, two-thirds received a favourable response, including agreement that the use of private cars for peak-hour journeys to the centre of large towns and cities should be limited, and that public transport should offer an attractive alternative. To accomplish the former, the Government agreed that parking-meter zones should be extended, and that, in future, liability to conviction for parking offences dealt with by fixed penalties, would fall absolutely on the registered owner of the vehicle, thereby closing an important loophole. To improve public transport, it was affirmed that revenue support to it was to be made more widely available in a proposed new structure of transport grants that the Government had mentioned during the later stages of the Select Committee's inquiry.

Precise details of the new grant system were released, after substantial discussion with local authority associations, in August 1973, eight months after the Committee's Report.[11] Under the new system most, though not all, of the specific grants (a grant for a new bus introduced in 1968 was a notable exception) were to be replaced by a unified system of block grants covering current, as well as capital, expenditure. The aim was to "eliminate the bias towards capital or current expenditure or towards particular forms of expenditure", thereby promoting a more comprehensive approach to transport and planning problems; and giving local authorities greater freedom for discretion in their spending programmes.

With the new system, part of the money previously expended on specific grants was to be absorbed into the rate support grant and the remainder distributed as a supplementary transport grant (TSG). The TSG was payable at a fixed rate to counties, because, with the reform of local government from April 1974, counties were made responsible for drawing up local transport plans. It was payable to counties on that proportion of their expenditure on local transport in a grant year that was both acceptable to the Government and exceeded a prescribed national threshold, based on a sum per head of county population. Whether or not proposed expenditure was acceptable to the Government was conditioned by two factors: first, whether the item of expenditure was on an eligible list, and second, whether the proposed spending was based on an acceptable "transport policy and programme" (TPP), which each county was required to submit on a yearly basis.

Though the announcement by the Department of the Environment in 1972 that provision for bus-operating subsidies was to be incorporated in the new system of transport block grants was, in itself, significant, the crucial indicator of the Government's conversion to a new approach was the *extent* to which it was willing to divert resources from urban road building. As the Select Committee had noted "what is required is a major and substantial shift of resources and management effort within the present and planned levels of transport programmes".

Evidence on this, too, was soon provided. By one of those curious coincidences that history produces from time to time, on the day following the tenth anniversary of Ernest Marples' unveiling of the Buchanan Report, *Traffic in Towns*, Marples' Tory successor,* John Peyton, announced to the House of Commons in a relatively short speech,[12] that he proposed a switch of resources in the transport sector away from urban road construction. He did not elaborate on the full significance of his decision, and, in fact, the uncertainties surrounding the imminent introduction of the new grant system (the enabling Local Government Bill had been introduced into Parliament only the month before) made a more informative statement difficult. But what the Ministerial statement did do was effectively confirm that the very basis of the post-Buchanan policy on traffic in towns was at last discarded.†

Summary and Conclusions

In Britain in the early seventies about two-dozen cities had populations greater than 200,000. Of these, around half a dozen had, by the early part of 1973, abandoned, or at least temporarily suspended, plans to build extensive highway systems. Although other cities, Glasgow, Liverpool and Leeds for example, pressed on with their own schemes with undiminished enthusiasm, the shock had registered. The received wisdom that had pushed so many cities down this road a decade before had at last witnessed a significant challenge.

The challenge had come in fact from a distinctly different approach to roads and traffic. This had developed in parallel with an increasing realization that there was a lower limit than Buchanan and his peers had assumed, to the amount of physical change to familiar surroundings that people would tolerate. As a consequence, it was an approach which stressed making better use of the existing road system and it had elements within it that could be traced back to both traffic engineering techniques first honed in Marple's days and to economists' arguments for marketing urban road space. (Indeed, one of the strongest propo-

*Marples was, of course, Minister of Transport whilst Peyton was Minister for Transport Industries under the Secretary of State for the Environment.

†Although the change in policy was both said and made to appear to be closely associated with a reassessment of the role of the railways in transport generally (and in a wider context by a need to divert resources to housing) in actual fact the Public Expenditure White Paper of December 1973 (Cmnd. 5519) showed that there were hardly any changes planned in resources going to urban roads: less expenditure on road construction was balanced by planned increases in spending on maintenance. But this goes to emphasize the significance of the change; it was not a change imposed by resource constraints. Equally the change had little to do with the petroleum crisis accompanying the Yom Kippur war of October 1973. The change of policy was preceded by a long term special Department review.[13] But, of course, the crisis substantiated the policy changes.

nents of the new approach, Michael Thomson, had been and continued to be an ardent advocate of road pricing.)

In the early 1970s this alternative was making considerable inroads into the middle ground of popular opinion, and by the end of 1972 had won the support of the bipartisan House of Commons Select Committee on Expenditure. By this time, the winds of change were evident to the Government too, and steps were being taken to accommodate the new mood. By the end of the following year intent had been transformed into action. A Local Government Bill had been introduced proposing a new system of block grants for transport purposes, which would allow for government-supported fare subsidies. More or less at the same time, the Minister for Transport Industries announced a shift of resources away from urban road building.

At the time, the full significance of these changes was not appreciated, lost as they were in the confusion of the oil crisis and its economic ramifications, and in the gestation period of financial reform for local government. But these pronounced, but possibly least recognized, U-turns of the 1970–74 Government of Edward Heath were significant indeed. Really they marked the end of a long struggle to maintain a semblance of the policy established before the war of operating urban bus services on strictly commercial lines. And they marked the end of the policy established a decade before, with the Government's acceptance in principle of the Buchanan Report, that when resources allowed, there should be substantial road building in British cities to maximize the degree of access for private cars. In fact with these shifts of policy, a distinctive chapter in the post-war history of traffic in towns was at last drawing to a close. Henceforth, after the autumn of 1973, the emphasis was to be quite different.

CHAPTER 10

Urban Traffic and Roads: Post-1973

BY the mid seventies the urban traffic policy that had applied for most of the previous decade had been stood on its head. This now-superseded post-Buchanan philosophy, put in simple terms, had been to build eventually as much road space as was physically possible and then to rely on traffic restraint and improved public transport to curtail any excess pressure. After Peyton's announcement in the autumn of 1973, the contrasting theme, succinctly expressed soon after in a circular to local authorities, was that "road improvements in urban areas should normally be provided to complement comprehensive traffic management or when demand cannot be satisfactorily met by traffic management and restraint measures alone".[1] In other words, new road space was no longer to form the lead item but to fall out as the residual element in a revised, and now somewhat complex, urban traffic policy that emphasized managing the demand for existing road facilities.

Demand Management: The Unexploited Potential

It was evident quite quickly that the Government believed there was still substantial scope for solving traffic problems without resorting to more radical measures. The traditional methods to be relied upon included better control of moving traffic, giving buses priority over other road users, and regulating the supply and price of parking space.

The Government had been trying to persuade local authorities to apply these approaches since the sixties with varying, but often little, success. Most success had come with controlling and assisting general traffic movement, and local councils on the whole had been active in introducing one-way schemes and on-street parking bans. The remaining potential here proved to be an unexpected bonus in applying computer control to traffic signals. The Government's Transport and Road Research Laboratory had been experimenting with this technology for some years and had used Glasgow and West London as a test bed. In Glasgow, average reductions in vehicle travelling time achieved, in comparison with manually prepared signal settings, was 16 per cent. In West London, journey times were reduced on average by 9 per cent, equivalent to a 6 per cent increase in street capacity. But, with the first purpose-built automated traffic control system for a whole city installed in Leicester only in 1974, clearly there was considerable mileage yet in Ernest Marples' favourite approach of squeezing the maximum amount of traffic through existing streets.

The comparative willingness of local authorities to apply traffic control measures did not extend, however, to bus priority schemes. Here the idea was that by introducing near-side lanes restricted to buses (and a few other selected classes of vehicle), the bus could bypass any congestion or get to the head of the queue at junctions. Alternatively, lanes permitting the bus to run against the flow of traffic (contra-flow lanes) allowed it to avoid more circuitous one-way systems. Restricting turns at junctions to buses only, also had the same effect. In this way the bus journey was speeded up relative to the car journey and, as a consequence, it was considered possible to "force the transfer of some of the people from cars on to buses".[2]

The idea had been mooted by Tom Fraser, then Minister of Transport, in 1965. Yet by 1970, a working group that the Transport Ministry had established to consider ways of easing the movement of buses around cities reported a

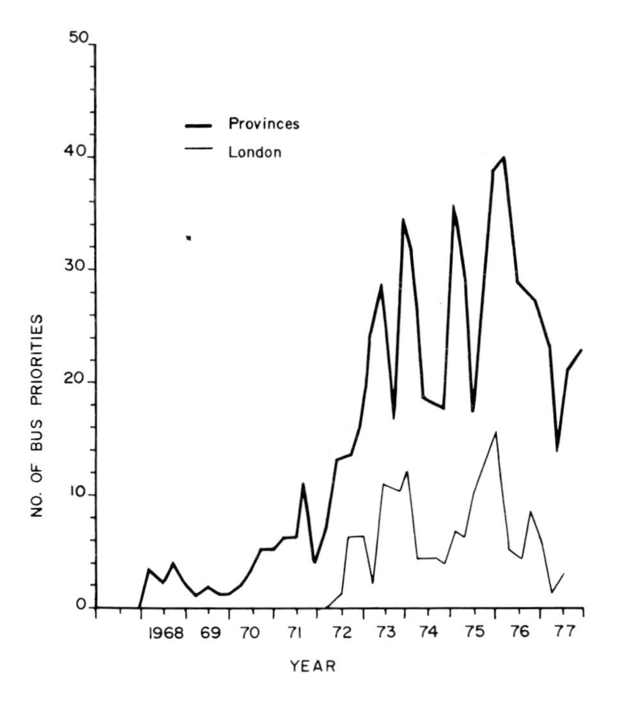

FIG. 10.1. Bus priority schemes, quarterly trends 1968–77 (third quarter). The boom came in the mid seventies with some evidence that the peak period of introduction had passed by 1977.

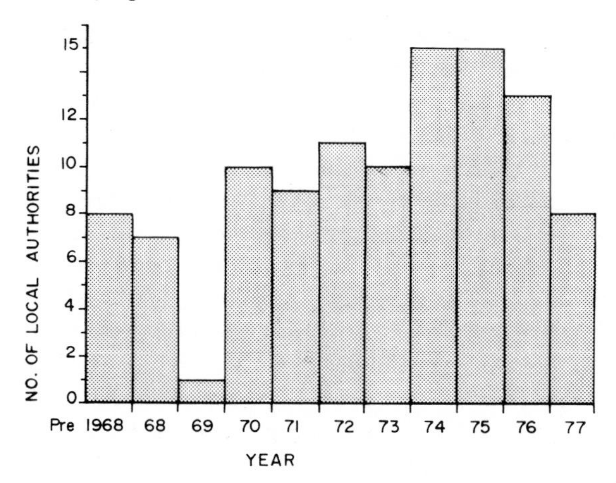

FIG. 10.2. Bus priority schemes, local authority initiatives: shown are the number of local authorities introducing schemes for the first time in a particular year. There were 107 for which definitive information on dates existed, although 120 authorities were known to have introduced schemes of one sort or another by late 1977.

general lack of interest and enterprise amongst operators and local authorities. In fact finding worthwhile schemes to monitor "involved a hard grind scouring the country . . . ".[3]

In the early seventies the picture began to change. In 1971 the Government introduced generous grants (increased in 1972) to cover the cost of signposting priority schemes and consequent road works. In addition the well publicized success of some early schemes in cutting bus running times and delays possibly engendered further interest. Even so, by the time urban road policy changed at the end of 1973, there still remained many opportunities for action. A large number of towns had still to introduce their first bus priority measures, and, even in those that had, the possibilities were far from exhausted.

The unexplored opportunities for parking restraint in the mid seventies were greater still. The Select Committee, in its 1973 Report, "Urban Transport Planning", had concluded that the regulation of parking was "weak, muddled and lacking in effectiveness and overall cohesion". That financial year, 1972/3, local authorities spent £48 million on covering the wages of parking attendants, on maintenance, and on paying debt charges on capital used for constructing car parks. In return they recovered only £32 million in revenue, leaving, in effect, a trading subsidy of £10 million.

In April 1975, spurred on by the advice of another Select Committee Report,[4] the Government decided to try to tighten up on the situation. They announced that normal deficits incurred on operating car parks would no longer be accepted as a claim under the new system of assisting local authority transport expenditure. As a result, the administration was soon able to note a welcome change in this situation, and a revenue surplus over and above operating costs was recorded from 1975/76 onwards. Over-all deficits on parking, including capital charges, also began to fall rapidly from a peak of just short of £20 million in 1974/75, but by the end of the decade there was still a loss of £4·5 million. In the conurbations, for example, only the GLC made a net profit, although South Yorkshire and Tyne–Wear Metropolitan Counties came very near to doing so.

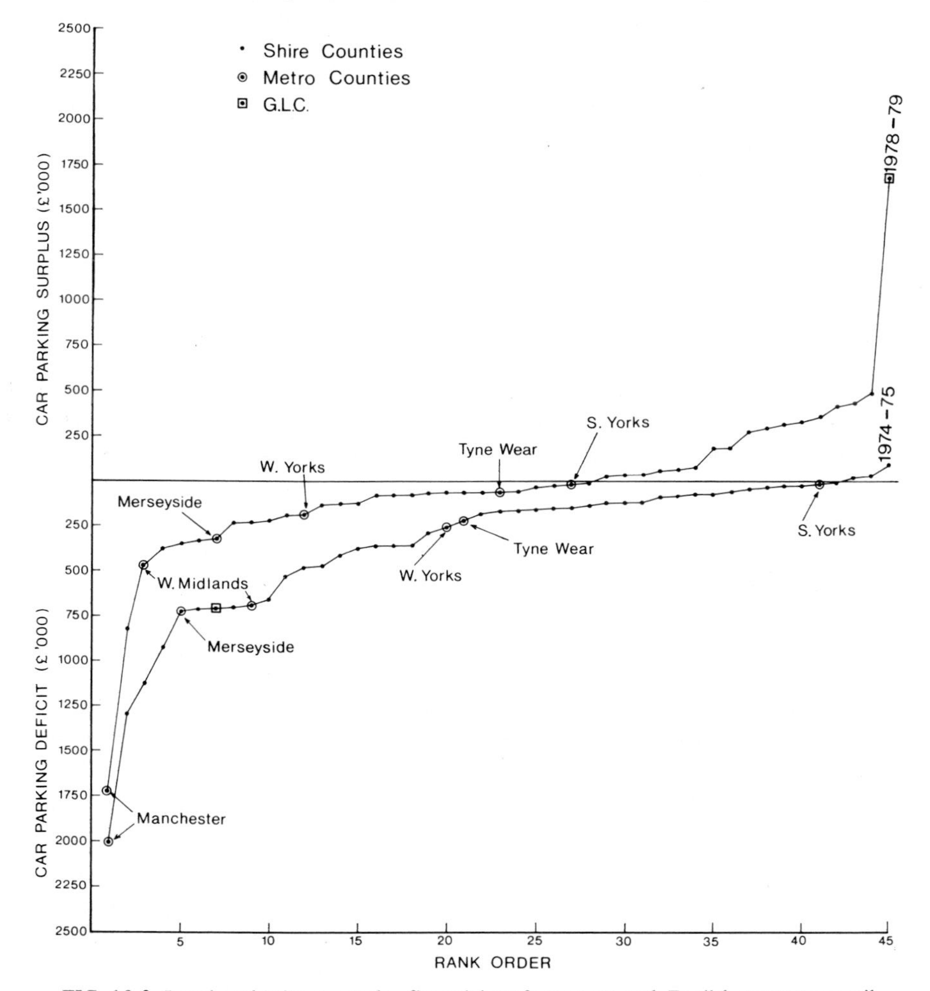

FIG. 10.3. Local authority car parks, financial performance: each English county council (except for the Isle of Wight) has been ranked according to its surplus or deficit on providing council car parking during the years 1974/75 and 1978/79. There was a general improvement during this period but by the latter date many authorities were still making losses.

The Government had also noted that District Councils, especially, were favouring in their planning consents a too generous provision of parking supply by the private sector. This was a particular cause for concern because, at the time, local authorities, except for the GLC, had no powers to control privately-operated public car parks or private parking under office buildings, shops and the like. Yet, as the Department of the Environment's 1976 Consultation Document on transport policy noted, these accounted for more than half of the parking space in many cities. The Government therefore announced in the mid seventies that it proposed to legislate to give English and Welsh Counties and Scottish Regional Councils powers for regulating the operation of privately-run public car parks. The 1978 Transport Act put this into effect.

The Government regarded this latter legislation as enabling local authorities to "widen their range of options to regulate traffic".[5] However, in the belief that considerable scope existed within the framework of the traditional approach to traffic restraint, the Government proved rather reluctant to widen the range of options too far. In fact, as was made clear in the run-up to the 1977 White Paper on Transport Policy, the Government remained far from convinced that the situation was such to justify what it rather pointedly referred to as "emergency action".[6]

The emergency action to which the Government was referring was of the kind that the GLC had been flirting with for some time. In 1965, soon after the Council was formed, preliminary discussions had taken place on the possibility of an experimental licensing system under which owners of private cars could drive into Central London only if they displayed on their windscreens daily licences or season tickets. This and related ideas were canvassed publicly in a GLC discussion paper titled *"Traffic and the Environment"* which came out in 1972, a time when the Council and the anti-motorway lobby were locked in a protracted battle during the Layfield enquiry. Layfield came down firmly in favour of introducing area licensing before the end of the decade. Thus stimulated, the Conservative administration at County Hall, almost as a last gasp before defeat at the May elections, launched a feasibility study of the idea. In 1974 the technical officers reported favourably but the new Labour Council decided not to take the scheme any further.

To get as far as implementation, the scheme would in any case have required legislation, and, as the 1976 Consultation Document and the ensuing White Paper revealed, the Government was far from keen on the idea. In the Consultation Document, the Government went to considerable lengths to argue its case, and, it might be added, in the process painted a fairly complacent picture of the traffic congestion problem. First, it pointed out that it was important not to lose sight of the fact that there was no point, on traffic grounds, in reducing congestion if the costs of doing so exceeded the traffic benefits of less congestion. The Consultation Document added that "clearly, if reductions in car traffic volumes were to produce insignificant improvements in journey times to other road users the case for traffic limitation would be weak". In effect, it then went on to argue that this was indeed the case.

The basic evidence used was data on vehicle speeds in different sizes of city, in central and non-central areas and at peak and off-peak times during the 1963–71 period (basically the information presented in Table 6.5). These data indicated that because of the limitations imposed by traffic lights, pedestrian crossings and so on, traffic speeds associated even with "very low levels of congestion can be expected to be quite low – almost certainly below 20 miles per hour in central areas". This suggested that the potential for improvement in peak-period speeds was only moderate. For example, a reduction of peak traffic volumes to levels similar to those in the off-peak (a time when "generally little serious congestion is in evidence") was thought to produce at the most a 25 per cent increase in speed.

Set against the prospect of only a limited improvement in speeds were certain costs and disadvantages which needed to be taken into account. For example, there were the so-called "disbenefits" to those motorists deterred from travelling as they had previously done, and the costs of accommodating extra journeys by other transport modes. In addition, there were the costs of setting up and operating the restraint system.

Evidence on the balance of advantage and costs was beginning to emerge at this time from two particular sources. One was a feasibility study of applying restraint systems to central Coventry carried out by a Department of the Environment working group.[7] The Coventry study analysed different systems including supplementary licences, and compared these with more rigorous parking control. It concluded that ' the systems would have high operating costs which, while not exceeding the benefits produced from congestion relief, meant that moving-vehicle controls would not be noticeably superior to parking controls at least until the late 1980s". It did, however, qualify its conclusion by pointing out that, because nearly two-thirds of parking spaces were in public ownership and therefore easy to control, Coventry was considered atypical.

Certainly the parking situation was quite different in Nottingham, where the essential features of the "zone and collar" proposal had been introduced on an experimental basis in the western part of the city in August 1975 (see p. 85). It soon became apparent that the scheme was not having the desired effect. There was negligible transfer to the bus.[8] One reason was that two-thirds of the traffic came from areas outside the control zones and thus had experienced no planned restraint until the inner collar was reached. But because of the layout of the road system it proved impossible to build-in very much delay at the collar controls without blocking other junctions. Consequently for most motorists, the general improvements in the public transport services *relative* to the car were rather small. Moreover in Nottingham's case, circumstances demanded that the relative change had to be large to be effective. These circumstances were that, in stark contrast to Coventry, about 90 per cent of Nottingham's car drivers travelling to work in the inner-city area were able to park free of charge. In addition, 25 per cent of those from the controlled zones received financial help from their employers for the running of their cars. Over-all, the economic calculations indicated that there were no benefits to commuters (but a cost of running the scheme of over £0·25 million). Therefore it was decided not to extend the experimental period when it expired in 1976.

Although by 1976 the Government had come down strongly in favour once more of parking controls supplemented by bus priority schemes, it did nevertheless recognize the Nottingham problem. Therefore it proposed in the June 1977 White Paper on Transport Policy to give County and Regional Councils powers to control private off-street parking. It was admitted that any such scheme would undoubtedly present administrative difficulties, but nevertheless an effective scheme was considered practicable. However, after consulting with local authority associations, the motoring lobby, the Confederation of British Industry (CBI) and the National Chamber of Trade the idea was dropped.

The administrative problems were unlikely to be eased or powers used effectively if the proposals were foisted on unwilling authorities, and too many authorities were proving unenthusiastic. The London Boroughs Association's Housing and Works Committee, for example, opposed the idea on the grounds that it was "unlikely to encourage industry or commerce".[9] Consequently, the range of options in the eighties for controlling parking was more restricted than anticipated in 1976/77. Whether this set-back advanced, in the Department of Transport's eyes, the date at which the more radical restraint measures might be required, remained to be clarified.

The Roads Residual

After 1973, road proposals became increasingly treated as a residual item, as the Government had intended. The Government pursued a number of approaches to ensure that this was indeed the case. At one extreme it simply offered advice, especially as part of the annual process of settling the Transport Supplementary Grant. Particularly notable in this context was advice given in 1975 which was very much in the spirit of a previous recommendation of the House of Commons Expenditure Committee that urban road building schemes that had not reached contract stage should be re-examined *ab initio*. Now local authorities were urged to review all major road schemes in their programme not due to start before April 1978, with a view to abandonment or reduction in scale.[10]

The urgency of such a review was soon emphasized when during the following year, in response to the serious balance of payments crisis and International Monetary Fund pressures, the Chancellor's cuts in public expenditure weighed heavily on roads spending. In fact a moratorium was placed on new road contracts between July 1976 and May 1977. But subsequently, with the lifting of this emergency measure, the Department of Transport advised councils that there would be very little scope for expansion, and the expectation was that local authority spending on road construction would continue at the (rather low) level planned for 1977/78.

This advice, of course, could be backed-up by firmer measures if these were required. As a last resort the Government could, and did, withhold approval for particular schemes under the planning control system.* For example, in 1975 Anthony Crosland, Secretary of State for the Environment, as if wishing to emphasize the new attitude to urban roads, refused to approve amendments to the Development Plan for York. These sought to provide an inner ring road, but Crosland was far from convinced of its necessity, and the council was asked to "explore more deeply further measures of traffic management and restraint and the greater use of public transport".[11]

In 1978 it was the turn of another historic centre, Oxford, and its notorious inner relief road, now proposed to run south of the confluence with the Thames and the Cherwell, to come under Government scrutiny. The further study of the Meadows relief road requested by Richard Crossman in 1966 (see p. 72) had been produced in 1968. It suggested amongst other things that the preferred route was south of Christ Church Meadow, and a version of this route was finally accepted by the City Council and the Department of the Environment in 1970, following a public inquiry. However, opposition developed (this time from householders rather than from the University) during the early 1970s, at a time when a policy of traffic restraint seemed to be gaining favour with the City Council. In these circumstances the Council reversed its policies, decided to abandon its road plans and to concentrate instead on a rigorous and comprehensive policy of traffic restraint and improvements to public transport. But this was far from the end of the matter. With the reform of local government, the County Council assumed responsibility for these matters, and the County, in its wisdom, reverted back to a policy incorporating a relief road. It was this road-orientated policy, forming a constituent part of the development plan for Oxford that, in the late seventies, was to receive the attention of the Secretary of State for the Environment. On this occasion, Peter Shore told Oxfordshire County Council that he intended to drop the scheme from the County Plan. Instead, he called for the preparation of new plans for the city based on "traffic management measures and the development of public transport".[12,13]

Thus in one way or another, urban road construction was forced into the hind seat. Expenditure on constructing and improving local roads, most of which was urban, fell dramatically. By the end of the seventies the level of spending on

*There was also another potentially powerful mechanism of control. Local authority capital expenditure is financed by loans. Borrowing for carrying out projects over £0·5 million (this is referred to as the "Key Sector") have to be sanctioned by the Government. However, there is no specific evidence of the Government failing to give Key Sector approval.

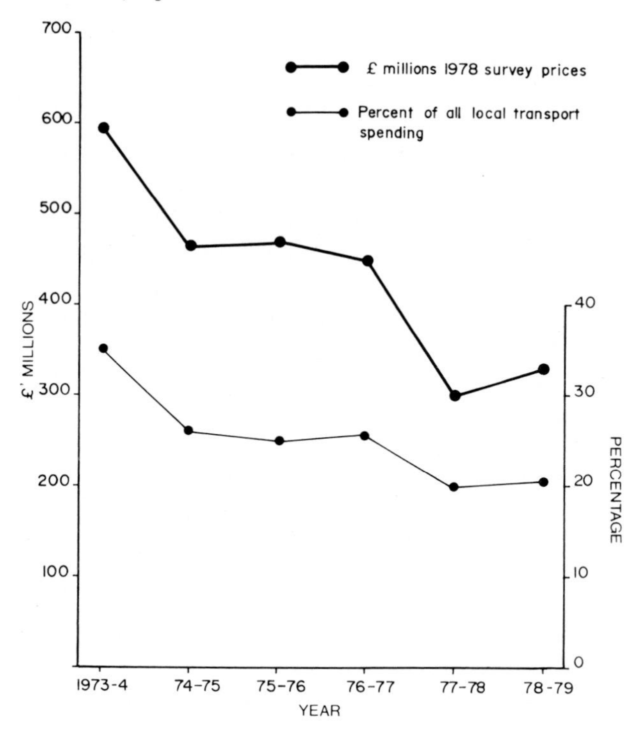

FIG. 10.4. Local road construction expenditure, 1973/74 to 1978/79: expenditure on new and improved roads fell both in absolute terms and as a proportion of all local transport spending after 1973/74.

new road works was little more than half of what it had been in 1973/74, and as a proportion of all expenditure on local transport it declined from just over a third to about a fifth.*

One casualty was the ambitious highway plans spawned by the transportation studies of the sixties and early seventies. By the end of the seventies most of these were well behind the schedule originally envisaged. In England, for example, after 1974 there was little growth in the mileage of motorways in the principal, and therefore largely urban, category. There were exceptions of course. Where an early start had been made and where the relationship with or dependence on motorways in the national trunk-road programme was strong, then the changes on the ground were often substantial. Glasgow was in the lead here with completion by 1980 of over a dozen miles of urban motorway

*On the basis of evidence[14] presented to a House of Commons Select Committee, in marked contrast to the programme of spending on trunk roads this fall does not appear to have been accentuated by failing to spend all the money set aside in the estimates.

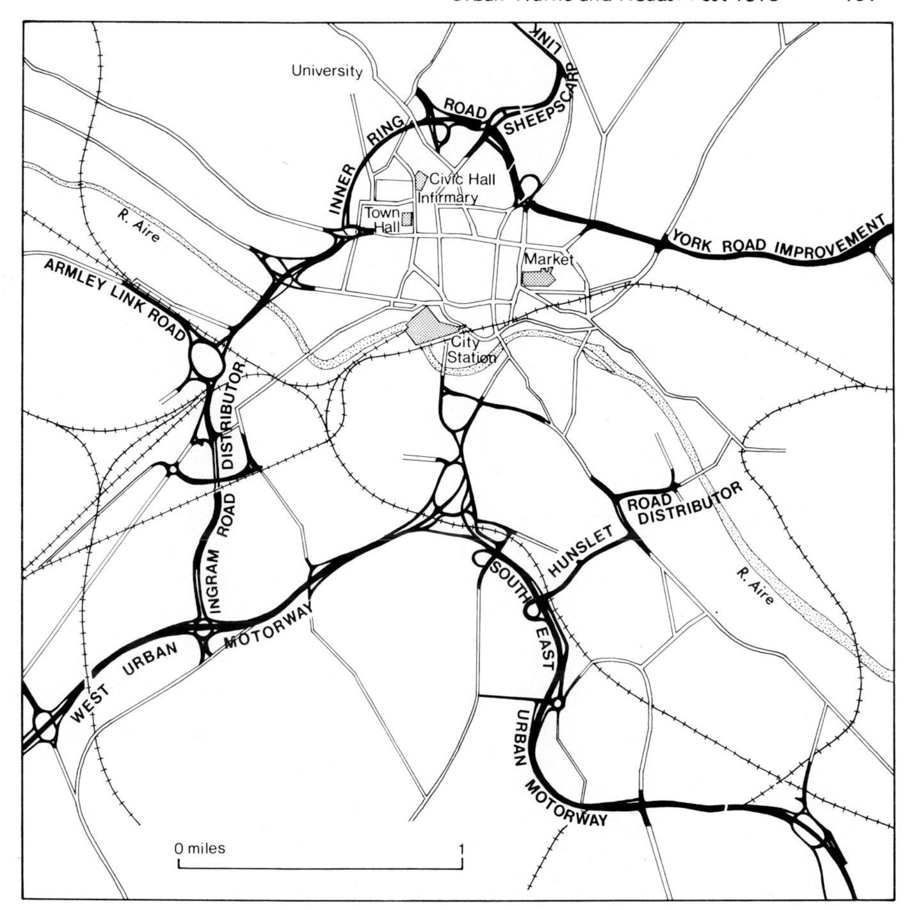

FIG. 10.5. Recent large-scale road development in Leeds: Leeds was one city to see its central area road plans substantially completed by the end of the 1970s.

linking in with the Strathclyde regional motorway network. A few English cities such as Newcastle, Liverpool and Leeds had also made an impressive start.

But it was more common for towns and cities to lie at the other extreme. By 1980, in London, where the slump in road spending outpaced the national fall, a section of the East Cross route north and south of the Blackwall tunnel and a small section of the West Cross route near White City were virtually the only parts of the notorious Rings to see the light of day. A similar picture prevailed

FIG. 10.6. Reading's incomplete inner distributor road: a further and final section was proposed running along the right-hand side of the photograph. But with new attitudes to urban road construction in the late seventies and less money available for road construction the proposal stalled and was finally axed in 1981. Vegetation soon covered the abandoned section in the bottom right-hand corner, known locally as the "ski ramp" (*see over*).

FIG. 10.6.

FIG. 10.6.

in cities which, like Nottingham, had been in the vanguard of the new traffic philosophies, and in historic towns like Bath and Cambridge, where environmental considerations were strong.

However, not only did the rate of implementation of the early road plans slump, but the actual schemes implemented were not always those originally given priority or even included in the original plans. At the same time as spending on new urban roads was being cut back, there were pressures to redirect what was left of this budget. Projects which fitted in well with comprehensive traffic management schemes or were capable of relieving the pressures of traffic on the environment now found favour.

In addition, during the 1970s there was an increasing belief that an inner-city problem existed. There was some disagreement as to whether it constituted a problem unique to the inner city, but the general focus was upon the untidy social, economic and aesthetic residue left behind in the wake of falling job opportunities and a declining population. It was not in fact until the fifties that the actual numbers living in the inner cities began to drop, although the proportion of people living there had been declining since 1931. But in the sixties and seventies, often encouraged by urban and regional planning policies, the exodus was dramatic. Between 1966 and 1976 Manchester lost 110,000 members of its population, Liverpool 150,000, Glasgow 205,000 and London 0·5 million.

In many ways the ingredients of the inner-city problems were considered similar to the long-recognized regional problem, and, not surprisingly, some of the proposed solutions were similar too. Here also it was felt that improved road communications could play a role in reversing the spiral of economic decline, and local authorities were asked to give priority in their road plans to inner-city projects. As the Deputy Secretary responsible for the Department of Transport's local road policies commented in 1980, "the major thrust" of urban road investment must be increasingly to assist these inner-city problem areas.[15]

There was, in fact, a certain ironic touch to this prescription. Some commentators[16] would claim that the 1960s plans for urban motorways, by encouraging the clearance of large areas which were subsequently blighted when the plans faltered, had made their own rather special contribution to the problem. In Liverpool, for example, as one astute observer[17] noted early in 1972, various pressures including road building had combined to produce vast areas of vacant land. Yet the extraordinary thing was that hardly anything was being put back. He found it ironic that the inner-city motorway had been shelved so that, as a consequence, the land acquired for it would remain unused.

There was, however, an even greater irony. The declining economy of the inner city had taken some of the steam out of the traditional argument for building inner-city motorways. As jobs and homes had disappeared, the pressure of congestion appeared to have eased too. But in the late seventies, in a curious reversal of roles, these same forces of change were on this occasion now being used to bolster the case for new roads. Indeed, in Liverpool the planners were now promoting an inner ring road (downgraded from a motorway) as an act of faith in an attempt to revitalize an ailing inner city.[18]

Summary and Conclusions

When the Secretary of State for the Environment gave his decision in 1975 on the application to amend the York Development Plan, he stressed that his general objective was the same as the Council's, namely to reduce substantially the volume of traffic in the centre of York. The City had proposed to do this by constructing an inner ring road. The Inspector at the public enquiry into the proposal concluded that the inner ring road was essential for the well-being of the city centre, for its conservation, for the establishment of an agreeable atmosphere in which the people of York and district would work and shop. The Secretary of State, however, came to a very different conclusion. He agreed that the proposed route for the new road was the least objectionable, but he was not convinced that all possible alternative ways of solving the traffic problem in the city centre had been explored.

This decision summed up the quite considerable change in attitudes and policy that, in a very short time in the middle years of the seventies, had demoted road building as a way of relieving towns and cities of their traffic burden. In many towns and cities less historic, larger and more industrialized than York, there was still considerable unexplored mileage in manipulating the demands placed upon existing road space, and the policy of the Government in the late seventies was to steer councils in this direction. Bus priority measures were encouraged, and the generous provision of cheap parking space for the car commuter greatly discouraged. But at the same time, the approach recommended was conservative in method and by degree: moderation was the essence.

It was as if the Government had suddenly discovered that traffic was largely self-regulating at a level of congestion that was tolerable and apparently not too inefficient. Indeed, by a strange irony, the Government began to justify its rejection of the more enterprising ideas for reducing city-centre congestion by adopting the same basic argument that local authorities had implicitly used in the sixties to disregard Government advice on traffic restraint. With speeds creeping up in spite of rising traffic, local councils at the time saw little cause to heed the Government's strictures on the need for restraint. Now for the very same reasons, it appeared that the Government was arguing that there was as yet no case for using what it had termed emergency measures such as supplementary licensing.

Meanwhile, expenditure on urban roads fell to much lower levels. It was also given new emphasis. It was related more to improving environmental conditions rather than simply raising traffic capacities and complementing traffic management schemes. As a completely new feature, it was directed at areas like London's Docklands, suffering as they did from inner-city decay, the idea shared with regional road planners being that new roads could help cure the malaise afflicting local economies.

CHAPTER 11

The Juggernaut Lorry Issue

FREIGHT vehicles had always been of relevance to road planning in post-war Britain. The 1946 tea room plan, for example, was heavily influenced by the need to move goods by road more easily in order to assist development areas and to boost the export drive. Then, at the beginning of the sixties, when the new "rolling programme" of trunk road improvement was getting into its stride, priority was given to those routes carrying the heaviest volume of industrial and commercial traffic. But it was in the seventies, largely on account of the juggernaut controversy, that lorries began to play an enhanced and especially important role in shaping the roads programme.

Forces of Change

The root-cause of the controversy was the very large *potential* economies afforded by the use of larger vehicles. Larger vehicles were, of course, more expensive to operate than smaller vehicles. Amongst other things, they required more fuel per mile, involved the use of more tyres and were more expensive to buy and to maintain. On the other hand, these costs rose more slowly than the vehicle's potential earning power as reflected in its carrying capacity. Consequently, when operating costs were expressed per ton of carrying capacity, costs fell dramatically with an increase in vehicle size. For example, in 1964 the ratio of costs per capacity-ton-mile for a vehicle capable of carrying five tons compared with an eight-ton-load vehicle, when averaged over 400 miles per week (the figures are from Harrison's mid-sixties study of road transport costs),[1] was of the order 14:11. Thus in this case, a simple transfer from the smaller to the larger vehicle would have resulted in a 20 per cent fall in costs, providing the proportionate load (the load factor) remained the same.

Of course, to realize these economies the crucial factor was an ability to carry freight around the country in larger loads. Clearly, in this vital respect the forces of change in the economy were operating in the desired direction and had been for some time. It was indeed proving increasingly possible to move freight in bigger and bigger loads. Why this should be so was not entirely clear. Possibly it was associated with the long historical process of industrial concentration into fewer and larger units of production; a process which was paralleled closely in the post-war years by similar trends in distribution and retailing. No doubt, also, this process of concentration was encouraged by faster and more reliable transit times associated with the developing system of motorways, and by a more flexible and competitive service following the abolition in the late sixties of controls regulating the capacity of the haulage industry.*

These developments which encouraged the use of larger-load-carrying vehicles were reinforced from the mid 1960s by radical changes in distribution systems associated with the introduction of container transport. The growth of containerization was an explosive feature of international trade in the 1960s, indicative of which was the fact that in 1962 only twenty-one ports in Britain had facilities for handling containers and unitized cargo. By 1969, however, this number had mushroomed to ninety-one. Of course, by 1969 total tonnages exported and imported, and therefore influenced by international trends in containerization were little more than 10 per cent of tonnages handled by inland transport. Nevertheless, the essential characteristic of containerization, namely its standardization for purposes of easy transfer between different modes of transport, meant that the influence of the concept on the road haulage industry as a whole was disproportionate.

Consequently, as likely container standards became evident once international discussions on the subject started in 1961, there were additional pressures to modify the "Construction and Use Regulations" that specified the maximum size and weight of vehicles. At this time the capacity of the largest permitted lorry of sixteen tons payload, and the maximum length of thirty-five feet, prevented carriage of all but an incompletely loaded international standard container of twenty feet in length. But in 1964 very substantial changes to both permitted weights and dimensions took place with a further minor change to maximum dimensions in 1968 (Table 11.1). The end result was that by 1968 lorries could legally carry an international standard container of twenty feet in length loaded to its maximum tare weight, and standard thirty- and forty-foot modules, albeit in these latter cases with some sacrifice of tonnages carried.

The consequences of all these forces of change were to be reflected quite quickly in the nation's vehicle fleets. The 1960s alone were to see a 45 per cent increase in the carrying capacity of the average vehicle, with, significantly, loads increasing by nearly the same degree. Amongst the heaviest class of vehicle, the growth in carrying capacities was even more startling. By 1970 the heaviest vehicles (those of more than eight tons unloaded weight) accounted for one-

*Following a report from the (Geddes) Committee on Carriers' Licensing in 1965,[2] the 1968 Transport Act phased out the system first introduced in the 1930s of A, B and C licences. The latter were usually available on request for businesses wishing to carry their *own* goods. A and B licences were necessary for professional haulage and existing carriers (including the railways) could object on various grounds, including the argument that granting the licence to the applicant would result in a capacity in excess of requirements. The 1968 Act substituted "Operators' Licences" available subject only to safety and good character requirements and resulted in probably the least regulated, most competitive road freight industry amongst Western economies.

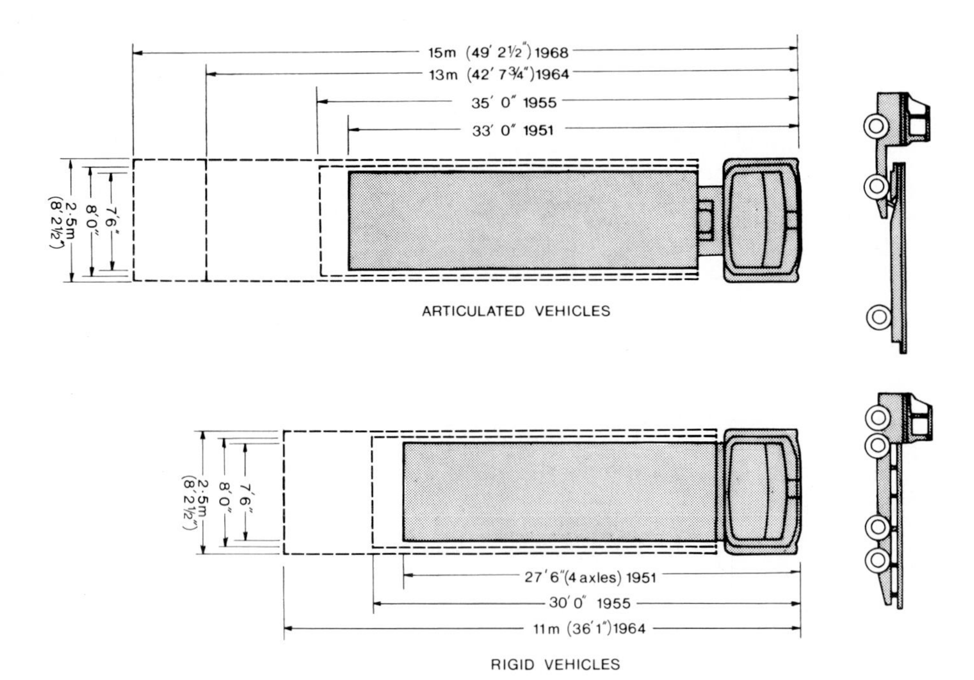

FIG. 11.1. Growth in lorry sizes 1951–68: the illustration shows the maximum size of articulated and rigid lorries allowed at various dates. The increase in permitted maximum dimensions during the 1960s was considerable.

TABLE 11.1. *Principal Changes in Weight Limits*

	Limits on gross weight (tons)				
	Rigid lorries			Articulated lorries	
Year introduced	2-axle	3-axle	4-axle	3-axle	4-axle
1955	14	20	24	20	24
1964	16	22	28	22	32
1966	–	–	–	24	–
1968	–	–	–	–	–
1972	–	24	30	–	–

TABLE 11.2. *Growth in Lorry Capacity and Output*

Date	Average carrying capacity (tons)	Tons carried per journey	Av. annual miles per lorry	Av. annual ton-miles per lorry
1960	4·67	2·99	16,400	43,000
1965	5·71	3·64	17,400	63,000
1969	6·79	4·19	18,600	77,900

Note: Ton-miles is a common measure of output obtained by multiplying tons carried per journey by vehicle mileage.

third of all ton-miles, three times the proportion of ten years earlier. This proportion was to virtually double again over the next six years.

In terms perceived by other road users and by bystanders, what these statistics really meant is vividly illustrated by commercial traffic on the M1 motorway. In 1966 about just over a tenth of the motorway's commercial traffic was of the largest four- or five-axle configuration. Seven years later this proportion was just short of one-third, and by the end of the seventies, taking the motorway network as a whole, the corresponding proportion was well over 40 per cent. The fact that, in time, larger lorries may have meant fewer lorries (total numbers stagnated after the mid sixties) was not appreciated. Quite simply big lorries were more than conspicuous and their growth in the 1960s was too gross and rapid to be assimilated easily. Thus, proposals to modify yet again the Construction and Use Regulations in the late sixties were seized upon as an opportunity to express a more general discontent with the environmental impact of large, heavy vehicles.

The 1969 Proposals

This time the proposals for change came from the SMMT, which advanced its case on the ground that it was impossible to meet overseas demand for high-payload vehicles competitively without the British home market supplying the volume base. Bearing in mind Britain's perennial balance of payments problems it was, of course, an argument expected to gain a wide measure of support. According to Kimber, Richardson and Brookes,[3] it had in fact been developed in consultation with Ministry of Transport officials, who had also guided the SMMT in formulating their proposal. But, whatever the merits and conviction behind the export argument, the potent force of containerization made its mark also on the proposal. Indeed, Mr. Mulley, when speaking to the House about the application for a change in the weight limits, referred to the possibility of accommodating fully-laden standard containers as being "the idea behind the proposals".[4]

The chief features of the proposal were: first, that the weight limit for five-axled articulated vehicles should be raised from thirty-two gross tons to forty-four tons (with modification to axle spacings); second, that for a lorry pulling a draw-bar trailer the limit would be raised to fifty-six tons (specifically to enable carriage to two fully laden standard twenty-foot containers); and third, that the maximum length should be eased half a metre to 15·5 metres. No modification to the all important ten-ton axle limit was requested.

In the middle of 1969 *The Sunday Times* decided to give coverage to these proposals, a significant event which was to mark the beginnings of a long-protracted battle between the freight industry at one extreme and the environmental lobbies at the other. However, one distinguishing characteristic of the objectors in this battle was the very wide range of representation. The roll-call included, of course, many organizations generally active in the transport field and especially in opposing motorways and other road building schemes – the Civic Trust, National Council for Inland Transport, the Pedestrians' Association, the Georgian Group, Victorian Society, Society for the Protection of Ancient Buildings, amongst many others. But, alongside these, also expressing various degrees of concern (largely on the grounds of presumed increases in road repair bills), were representatives of various and influential branches of local government. These included the Association of Municipal Corporations, the County Councils Association, the National Council of Parish Councils and the Urban and Rural District Councils' Association.

Both these broad groups of "objectors" lobbied intensively during the autumn of 1969 and were to see some return for their efforts during the debate in both Houses in December, when the Government was made well aware of the strength and breadth of feeling on the issue. Possibly in response to this show of strength by the opposition, Mr. Mulley, the Minister of Transport, promised to consult many interests before authorizing any change in the regulations. As a first step in this consultation procedure a private meeting of the protagonists (including representatives of all the local authority associations, the chief amenity bodies, the planning and architectural professions and the motoring lobbies) was convened in February of 1970. Thereafter, consultations proceeded for most of the year.

During this period a notable role was played by the Civic Trust which had solicited 700 local civic societies prior to the February meeting and followed this up with a well researched, well written and well presented memorandum condemning the heavy lorry on environmental grounds.[5] The memorandum was presented to the Minister in September 1970 and released for public consumption at a press conference on 6 December. Its effect was to reinvigorate the campaign at a crucial stage.

That summer the Conservatives had won an election with a manifesto promising to give greater emphasis to safeguarding and improving the environment. Now the new Government was under considerable pressure to act upon that promise by rejecting the idea of heavier lorries. The campaign paid dividends. Within a week of the Civic Trust's press conference in December, Mr. Peyton, the new Minister for Transport Industries in the new Department of the Environment, announced that the government did not intend to proceed with the SMMT's proposals. It appeared as though the environment lobby could claim an important victory.

The European Dimension

Within a month of Peyton's announcement, however, the issue of lorry weights was to move into an altogether different and more significant arena – that of the European Economic Community. This was the outcome of the Conservatives' European policy and its commitment to take Britain into Europe. One consequence of this move was that Britain (together with the other two applicant countries, Ireland and Denmark) was now party to discussions on new policy initiatives in the Common Market. Of these initiatives one was a proposal to harmonize lorry weights in accordance with a clause in the Treaty of Rome which called for the formation of a common transport policy. Ironically, draft regulations recommending a gross vehicle weight of forty-two metric tonnes and an axle weight of 11·5 metric tonnes (the latter well in excess of the

"Smash the next lamp on the left, flatten the pavement by the pub, nudge the sweet shop, scrape the Market Cross, then just follow the skid marks to London."

FIG. 11.2. A comment by *Punch* on juggernaut lorries.

British industry proposals) was presented by the European Commission to the Council of Ministers for their consideration within months of Peyton's announcement not to proceed with increased lorry weights.

Fortunately for the Government, the existing EEC members were unable to agree fully amongst themselves until May 1972, when agreement in principle to a forty-tonne gross vehicle weight and an eleven-tonne axle weight was reached. The intention was that these limits would come into effect in 1980, with existing national limits continuing for domestic traffic until 1985.

However, at this stage, negotiations for Britain to enter the Community had progressed to such an extent that a binding agreement was deferred. By the time of the last meeting of the Council of Ministers before Britain formally joined the EEC on 1 January 1973, the Commons had debated already and at some length an Opposition motion "that this House, mindful of the environment, is against bigger and heavier lorries".[6] The motion was moved by Anthony Crosland, who was later to become Secretary of State for the Environment with responsibilities which included transport. In an eloquent and well informed speech he reminded members that the issue was one of balancing economic and environmental factors and, in his view, in this instance, the balance was in danger of tilting too far against the environment. With two exceptions all the speakers in the debate shared this sentiment and often expressed vehemently their shared belief. As Angus Maude pointed out, this, together with the strength of public feeling on the issue, could not fail to strengthen the Minister's hand in negotiation with Britain's soon-to-be partners in Europe. The Minister, for his part, made passing reference to the benefits the export trade would gain from harmonization and left open the possibility of larger vehicles "only for the most powerful reasons and subject to strict limitation as to routes". But he added that he could not, without denying all the arguments he had used in Europe, do other than accept the motion on behalf of the Government.

After Britain's membership of the EEC, Ministers continued to argue strongly Britain's case. For example, in 1976 when the EEC Council of Ministers again considered the question of increasing maximum permitted lorry weights within the community, William Rodgers, the new Secretary of State for Transport (a Department of Transport separate from the Department of the Environment had been formed in 1976), made it clear that he would not consider such an increase in the UK unless he could be satisfied that there would be no resulting adverse effects on road safety or on the environment.

In January 1979 the EEC Commission's latest in a long line of proposals on lorry weights was published, this time in the form of a draft Council directive. Partly to forestall this European pressure, William Rodgers announced that he was considering a major independent enquiry to look into the economic and environmental effects of heavy lorries. In March, only days before Parliament was dissolved for the election which brought Mrs. Thatcher to power, Rodgers confirmed that he had decided to go ahead. In the event, it was left to Norman

Fowler, the new Conservative Administration's Minister for Transport, to make the final arrangements.

The Armitage Inquiry

The arrangements in fact were slightly different from those indicated by the Rodgers statement. In July, Fowler appointed Sir Arthur Armitage, then Vice-Chancellor of Manchester University, to undertake the inquiry. He was assisted by four assessors. These assessors included another vice-chancellor and two professors (with expertise in engineering, environmental medicine and economics), and the County Planning Officer of Merseyside County Council. The inquiry began in September 1979.

Although the terms of reference were broad and the inquiry was to be wide ranging, the spotlight and, consequently, most of the evidence focused inevitably on the issue of heavier lorries. The basic case of the juggernaut opponents, well represented for example in an extensively researched submission, *Weighing the Evidence,*[7] from the pressure group Transport 2000, was based on three main strands.

First, they argued that the economic case for having larger lorries was weak. The operating costs per capacity ton of the proposed juggernauts would be no less, and possibly more, than for the existing heavy lorries. Some support for this view came from the Government's Transport and Road Research Laboratory (TRRL), which examined operating costs for currently used vehicles and found that "while there are considerable savings in carrying more goods on heavier goods vehicles up to about 24 tonnes [gross vehicle weight], the additional savings from using even heavier vehicles are very much less and the cost per tonne-mile levels out beyond 26 tonnes implying that the savings are small for heavier vehicles".[8]

Second, it was pointed out that at least for the heaviest vehicles the utilization of the vehicles' weight capacity had been falling in the seventies. This view was also supported by the TRRL. Consequently, the argument of having lorries capable of carrying heavier loads was becoming weaker and indeed was reducing the economic case still further. Heavier lorries were only cheaper to operate if they carried heavier loads.

Third, the anti-juggernaut lobby pointed out that one startling feature of the seventies was that goods were being carried on average much further, in fact about 40 per cent further in 1977 as compared with ten years earlier. Admittedly, there were many factors that contributed to this, notably changes in the structure of industry. However, it was argued that the economics of larger lorries had been a powerful force in bringing about these changes. Therefore, the point made was that *if* still larger vehicles were more economic, then the reduction in traffic due to fewer juggernauts being needed for the same amount of work would be nullified in whole or part by there being more work to do solely on account of the larger lorry.

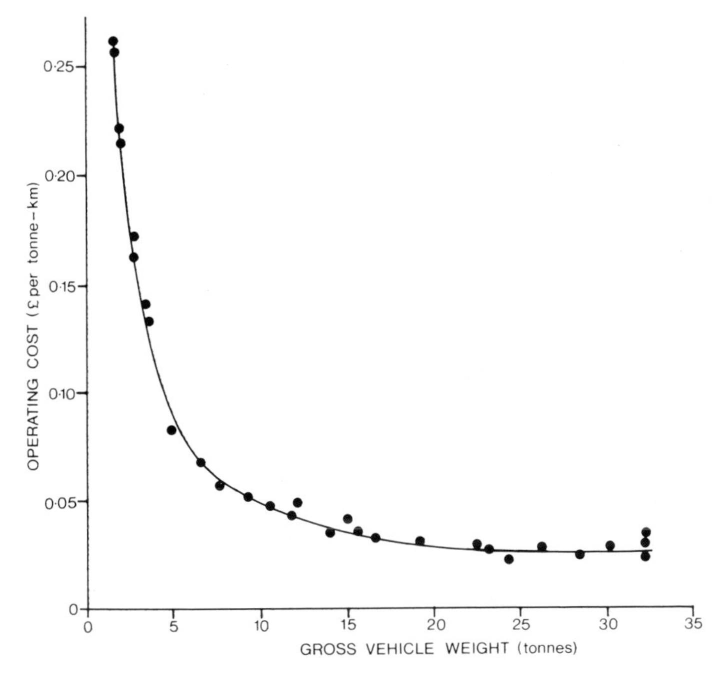

FIG. 11.3. The economies of large lorries: costs per tonne-kilometer fall rapidly with vehicle size up to about twenty tonnes gross vehicle weight. Thereafter, savings level out and appear negligible beyond thirty tonnes. However, it was assumed for purposes of the calculations that all vehicles were operated in the same manner with maximum loads.

Armitage presented his report to the Minister just prior to Christmas 1980.[9] On the issue of whether there should be still heavier lorries he made a number of points. First, he concluded that by having the main types of heavier lorry proposed by the EEC, costs per tonne-mile would be 13 per cent less than for the heaviest of existing vehicles. This conclusion was based largely on estimates prepared by "the industry" and by the TRRL, and on an assumption (contrary to recent trends) that load factors would increase slightly. With added qualifications these economies were considered likely to reduce the nation's freight bill by between £100 million and £160 million a year (at mid-1980 prices). Armitage added, without quantifying the point, that "much of the benefit should feed through to prices to the consumer".*

Second, Armitage believed that there would be environmental benefits

*However, as the savings represent about 1 per cent of annual expenditure on road freight, which in turn represents about 10 per cent of total costs to industry and commerce, the effect on final consumer prices would be about 0·1 per cent or the equivalent to three days' inflation at the rate current at the time.

FIG. 11.4. One argument used in favour of heavier lorries: the four four-ton capacity lorries in the photograph carry the same weight of goods as the two eight-ton lorries or the one sixteen-ton lorry. The juggernaut proponents argued on these lines that heavier lorries meant fewer lorries.

because heavier lorries each carrying more would mean fewer lorries; it was thought that approximately 8·5 per cent of the 1978 mileage by lorries over twenty-five tonnes would be cut. The generation of new demand for road transport as a result of the cost savings of heavier lorries "would be likely to be small" (he did not elaborate further), and, to allow for this, the initial estimates of mileage saved by having heavier lorries was cut back half a per cent to 8 per cent.*

Third, if more stringent standards on noise and braking distances were introduced (the Report suggested what these might be), Sir Arthur was of the view that the superjuggernaut could be both quieter and safer. Moreover, a reduction in heavy-vehicle mileage on account of heavier lorries would also lead to fewer accidents.

Finally, on the question of damage to roads and bridges, it was conceded that the EEC proposals on axle weights could have serious repercussions, particularly on the cost of repairing and rebuilding bridges.

Taking into account all these points, the inquiry's chief recommendation was that heavier lorries should be allowed up to a gross weight of forty-four tonnes with axle combinations and gross weights similar to those in the draft Council directive of the EEC. However, to avoid the bridge and road damage problem, the axle weights recommended by Armitage deviated rather more than did gross weights from the EEC proposals, particularly with respect to the axle through which the vehicle was driven. For this a limit of 10·5 tonnes, up 3 per cent on the existing 10·17 tonnes, was suggested. It was also recommended that maximum lengths of articulated lorries should be extended half a metre to 15·5 metres (specifically to allow improved design of the driving cab), and that a height limit should be introduced.

Conclusions

The juggernaut lorry controversy provides perhaps the most concrete example of the power of environmental lobbies to cast their mark on policy. For over a decade the lobby managed successfully to block powerful industrial pressures for an increase in the gross weight and dimensions of goods vehicles. It would appear that this success was simply due to the environment being an issue with sufficient political punch. In October 1978, for example, *The Times* reported that the then Secretary of State for Transport accepted the logic of the industry's case but that he always concluded meetings with the haulage and motor industry

*This was probably the weakest element in Armitage's case. To know the balance of advantage between costs and environmental impact, it was important to confront the issue of how much juggernaut traffic would be generated by reduced costs. As the Department of Transport remarked in their evidence to Armitage[10] "estimates of the balance of effects depends heavily on assumptions made about generated traffic, which in the long run includes the results of relocation of industry or alteration of distribution methods" (paragraph 6.15). Armitage did not call for any incisive econometric analysis of this issue.

FIG. 11.5. A thirty-eight-tonne heavy lorry: the photograph shows "Road-train", a new design by Leyland launched in the spring of 1980. It was designed to operate at thirty-eight-tonnes gross vehicle weight, in countries where this was legally permitted. In Britain, of course, it could operate only at thirty-two-tons.

delegation by telling them that it was not politically possible to do as they wished because the anti-truck lobby was too strong. Quite simply, at the end of the day it was a matter of relative power; in this contest the anti-juggernaut lobby proved more powerful politically than the pro-lobby.

The situation was helped, of course, by an industry failing to get its message across. Perhaps it was at times too preoccupied with other issues such as tachographs and changing regulations on hours of work to organize a successful case aimed at a wider public. Not until 1979 with the publication of *Freight Facts*[11] by the Freight Transport Association did this situation change to any marked degree. Not surprisingly, the industry frequently got a poor and sometimes, as in the case of *The Sunday Times,* a hostile press. There were exceptions, notable amongst which were two pieces in *The Times* in October 1978. On both occasions Michael Baily (the transport correspondent) and Clifford Webb (the industrial correspondent) raised an issue that Leslie Huckfield (the Transport and General Workers Union's parliamentary representative) had mentioned in the House of Commons debate on lorry sizes back in 1972. On that occasion Huckfield questioned whether members were "undergoing some kind of cruel deception". The point he made was that the forty-ton lorry was no larger physically than a thirty-two-ton lorry. In fact, as Webb had pointed out, lorries capable of operating at more than thirty-two tons had been running around – partly loaded – on British roads for many years. It had been added that providing the forty-ton lorry had more wheels to spread a load, such a vehicle probably did no more damage, but by carrying extra freight it might reduce the total number of large lorries on the roads.

It was these same fundamental points that were to be repeated with some differences of emphasis, eight years later by Sir Arthur Armitage when making his recommendations to the Minister to allow heavier lorries up to forty-four tonnes on British roads. In certain respects his supporting evidence was far from convincing – particularly on the issue of generated lorry traffic – but in some ways it was perhaps now of less importance. By this time the relevant environment had changed. Britain had suffered eight years of punitively low economic growth, and in the late seventies had experienced de-industrialization at the expense of a strong petro-currency. With over 2 million unemployed, manufacturing industry under great stress and road hauliers facing their worst recession for five years, proposals which promised to cut industry's costs and make the state-aided British Leyland commercial vehicle division more competitive overseas were bound to carry more weight. As the Local Government Operational Research Unit perceptively remarked in a report published on heavy lorries only days before Armitage reported, "environmental problems become less significant in periods of economic hardship".[12]

It might also be added that although the lobby was successful in blocking the proposed increases in maximum gross weights and dimensions the average size of a heavy lorry on the road and the loads it carried continued to grow. Statistics presented in the Foster Committee's Report on Road Haulage Operators Licensing[13] suggested that the same basic economic forces that prevailed in the fifties and sixties continued to exert their influence in the 1970s. The average carrying capacity of vehicles of over 3·5 tonnes gross weight increased between 1973 and 1976 by over 8 per cent, with loads almost keeping pace.

Thus in some senses, the environmental victory on the issue of larger juggernauts was a hollow victory. The balance between economic and environmental factors did not really change in a meaningful sense. For it to have done so, the environmental lobby would have had to take from "freedoms" that already existed. This, as the issue of lorry routeing was to show, was a proposition very different from that of denying what the road transport industry did not already enjoy.

CHAPTER 12

Lorry Routes and Road Plans*

IN 1964 the local council at St. Albans, Hertfordshire, approached the Ministry of Transport with a request for a lorry-ban to be placed on the A5 and A6 trunk roads passing through the town. When submitting their case, the councillors drew attention to similar successful bans operating already in other historic towns such as Gloucester and Oxford.

Ernest Marples rejected the council's request.[1] His grounds for doing so were that in Oxford and Gloucester the alternative routes were along ring roads passing close to the town centres: in St. Albans the only practical alternative (along the M1/M10 motorways) would have involved a considerable detour. He did, however, offer to alter the signposts in the hope that the lorry drivers would be attracted without compulsion to the motorways.

Neither the council's submission nor the Minister's decision was of any particular consequence; the case did not become a *cause célèbre*. It was merely one of countless requests that the Ministry received as a matter of course from local authorities for traffic orders to be placed on trunk roads. But what it did contain within its own rather parochial setting was the basic ingredients of an issue that was to be enlarged upon at the national level nearly a decade later. These ingredients were: first, the Ministry's predilection for self-imposed restraint on the part of the haulage industry; second, the Ministry's reluctance to impose too great a commercial penalty in deference to environmental gains; and third, the crucial importance attached by the Ministry to there being suitable alternative routes for lorries to use.

The Government's Strategy

In the late sixties the trunk-road master plan established by the Tory Government at the beginning of the decade was well on its way to completion. Harold Wilson's Labour Government had chosen to maintain the momentum of spending on trunk roads – both motorways and dual carriageways – and by 1968 close on 750 miles of the former were in use or under construction. With prospects

*Parts of this account are based on an unpublished paper ("Road Building and Lorry Restraint Policy, 1969–1977") kindly made available by Philip Blake of the Local Government Operational Research Unit.

of the target mileage being fulfilled in the early seventies, the administration turned its attention to the next stage.

Consequently, in March 1969 a Green Paper, *Roads for the Future*, commented on the imminent achievement of the 1000-mile target and added, "we need a new strategy for building and improving national roads in the 1970s and early 1980s".[2] What was proposed by way of new strategy was an extended high-quality network (shown indicatively on a series of regional maps) costing, at prices then current, about £1·6 billion with a further £0·6 billion to be spent on link roads and local improvements.

The sequel to this 1969 Consultation Paper was a White Paper published in May 1970.[3] Its title was the same, and its basic content was much the same, as that of the Green Paper. The only differences in the White Paper were that the basic network was slightly different from and larger than that portrayed a year previously. It was to cost £2 billion at 1970 prices, about £250 million more than the Green Paper's proposals. In concrete terms, this meant a high-quality 4500-mile network of dual-carriageway roads completed within fifteen to twenty years.

But six months later when the new Government led by Edward Heath was formed, these White Paper proposals were back in the melting pot. This time the review of road plans was part of a wider reappraisal aimed at reducing the over-all role of government. The idea was to cut the burden of taxation by reining-back public expenditure.

The roads review was completed and ready for unveiling by the middle of 1971. But in spite of the new political philosophy of the time, there was little of real substance to distinguish the outcome from the earlier Labour proposals. Surprisingly, the rate of spending was maintained (although there were cuts, albeit small, elsewhere in the transport programme). There was, however, some window-dressing to give the appearance of change. The target mileage of the earlier White Paper was repackaged in terms of 3500 miles within ten years.

The window-dressing also extended to stressing that the objectives of the new road-building programme included "diverting long-distance traffic, particularly heavy goods vehicles from a large number of towns and villages . . . ".[4] This emphasis had been prompted by the initial shots in the anti-juggernaut campaign of the environmental lobby. When proposals for lorries in excess of forty tons first surfaced at the end of the sixties, one response of the environmentalists had been to suggest that heavy lorries be restricted to a designated network of routes. In February 1970, for example, the Civic Trust, responding to the proposal for heavier lorries, submitted to the Ministry a memorandum noting that:

> "There seems to be real objection to present loads being allowed on almost any part of the road network, let alone the increases proposed. It seems essential that the heaviest commercial traffic should be restricted

> to designated lorry routes and prevented from passing through residential streets, town centres, conservation areas and minor rural roads."[5]

Later that year, the Trust's case was corroborated with the publication of the Report of the Marshall Committee, set up to consider more effective methods for maintaining highways. The Committee's Report argued that the use of unsuitable roads by heavy lorries added considerably to the road maintenance bill, and it recommended that the Government consider the possibility of national legislation to confine heavy vehicles to major roads.

The Government's reaction to these proposals was very much in keeping with the decision on the St. Albans request handed down by Ernest Marples six years earlier. The feeling was, as Mr. Peyton made clear in answer to a Parliamentary Question in November 1971, that legislation restricting the routeing of lorries would " . . . add intolerably to the costs of transport".[6] Nevertheless, the Government was well aware of the problem, which it felt could be resolved in two ways.

First, it aimed to make the heavy lorry "more civilized".[7] It aimed to do this by reducing the maximum noise levels allowed for new vehicles, by imposing smoke controls on new diesel-engined vehicles, and by introducing a minimum power-to-weight ratio for goods vehicles. Draft regulations on these matters were released in December 1970 and, after further consultation with manufacturers and operators, firm measures were announced in the autumn of 1971.

But the keystone of the Government's approach, and the element to which it attached most importance, was a modernized trunk-road network. Indeed, as the Transport Minister had conceded "as we get nearer to a comprehensive network of good roads . . . it might be possible to . . . restrict . . . heavier vehicles to use those roads which can accommodate them" (and at the same time permit still heavier vehicles).[8] However, and this was the very essence of the Government's case, suitable roads must come first.

With the Conservative Government's roads policy established by the middle of 1971, John Peyton, the Minister for Transport Industries, tried in the following months to hammer home the Government's basic strategy on the heavy lorry and to stress the importance of a high-class network of primary roads. In a speech to road hauliers on 9 May 1972, Peyton claimed that:

> 'A first-class road system was the top priority . . . As the road network becomes more adequate to our needs, so it will become easier to keep vehicles out of places where there is neither room nor reason for them to be."[9]

Two weeks later, speaking in Kent, a county in the front line of a disturbing invasion by juggernauts from Europe, he added that:

> "The large vehicle has no place in narrow city streets or quiet country lanes; its need to pass through towns and villages will grow steadily less

as the road system develops. Not the least of the advantages of the growing modern road network will be the greater possibility of protecting such places from unwarrantable intrusion."[10]

The message was emphasized again four months later. In a speech this time to the RAC, Peyton (referring to lorries) said that:

"Restrictions alone cannot be the answer to our problem; there must also be facilities which make restrictions sensible and acceptable."[11]

In other words, fulfilment of the inter-urban road strategy was seen by the Heath Government as an essential precursor to any attempts at restricting heavy lorries to a designated network.

But many environmentalists were far from convinced of the efficacy of such a strategy. Their view was that the Government was concerned too much with longer-term solutions. In the short-term it was felt that it gave undue weight to commercial factors at the expense of the environment.

The Dykes Act

This tide of concern for the environment was used in 1972 by Hugh Dykes, MP, to move a private Member's Bill seeking to regulate and control the movement of heavy commercial vehicles. Private Member's legislation requires sponsors. Indicative perhaps of the ground-swell of feeling on the matter, sponsorship of Dykes' Bill was by members of all major parties consisting of seven Conservatives, one Liberal (David Steel) and four Labour sponsors.

The Bill proposed that an Advisory Council be established to advise the Secretary of State for the Environment on the location of "red zones" where heavy lorries would be prohibited and "amber zones" where their use would be restricted.* If the Secretary of State accepted the Council's proposals, modified or unmodified, then local authorities had to introduce appropriate traffic orders. To strengthen their hand, alterations to the pertinent 1967 Road Traffic Regulation Act were proposed, explicitly introducing amenity as a basis for restricting the movement of heavy commercial vehicles.

Although the Bill as a whole received virtually unanimous support in the House, certain clauses within it did prove contentious. For example the introduction of amenity as specific grounds for restriction was argued to be unnecessary. The 1967 Road Traffic Act already gave powers to local authorities "for preventing the use of roads by vehicle traffic of a kind which . . . is unsuitable

*It is interesting to note the parallel between Dykes' idea and the proposal of a speaker at a road engineers' conference in 1964. The engineer canvassed a proposal for a nationwide system of lorry routes. His idea was that different weight-classes of lorry would be colour-coded and restricted to roads marked with the same colour.[12]

having regard to the existing character of the road or adjoining property".

But this aspect apart, in the Government's view (and that of the local government organizations) the proposed legislation had "one severe disadvantage", to wit the proposed Advisory Council removed responsibilities from local authorities, where the Government considered that they "should properly lie" (*sic*).[13]

Eventually, the Bill was redrafted and passed into law, with county councils replacing the Advisory Council in the initiating role. Consequently, a statutory duty was placed on local authorities to survey their areas, bring forward their own proposals and, by 1 January 1977 (1978 in Scotland), to introduce at least one draft order. The caveat was added that the county councils would be obliged to consult each other during the formation of their plans. Informally, however, Keith Speed, the Under Secretary of State, had given strong indication that the regional offices of the Department of the Environment would endeavour to provide a mechanism of co-ordination. In fact, prior to Dykes introducing his Bill, the Government had taken steps already along these lines. Through the regional offices discussions had started with local authorities on the possibilities of a system of *advisory* lorry routes related to the primary road network. But after Dykes' legislation had received the Royal Assent, these exploratory discussions were emphasized and provided with a focus.

The Rise and Fall of the Lorry Network

In July 1974, within months of the Labour Government's winning the "Miners' Election", an important development came with the release of a consultation paper titled *Routes for Heavy Lorries.*[14] What the paper proposed was that a system of routes should be set up in England focused on the network of high-quality roads expected by the early 1980s. This system, as it was clearly stated in the Minister's foreword to the paper, "would also provide a framework for the lorry plans which local authorities have to prepare under the Heavy Commercial Vehicle (Controls and Regulations) Act 1973". The paper itself went on to add that:

> "The national network will provide the most obvious routes (and most economical in operating terms) for lorry drivers between the main towns. Making its use, . . . mandatory would be to add a sanction to secure what the majority of drivers would do anyway. The main problem will be to control the movement of heavy lorries away from the national network."

This, of course, was where the Dykes Act came into the picture. But in spite of the change of government and in spite of the Dykes legislation, the basic message was still that spelt out by Mr. Peyton in 1971, namely, once a network of good routes existed, then one could think in terms of mandatory restrictions.

Of course, the inherent paradox was, as the above quotation comes close to admitting, that once a desirable network exists then mandatory restrictions are no longer necessary.

However, in 1974 achievement of a desirable network was far from resolved. In fact at the time that the Consultation paper was released, the Government's road plans (and therefore its lorry strategy) were beginning to fall apart as the trunk-roads programme was trimmed in response to cuts in public expenditure. The first cuts had come in May 1973, just six weeks after the third successful reading of the Dykes Bill. More followed later that year and the 3500-mile target set out in June 1971 receded to the mid 1980s.

In June 1974 in the face of yet more trimming, the Department attempted to salvage the situation. It had now been discovered that because of improvements in the performance of drivers and vehicles, roads could carry more traffic for the same quality of journey than as at one time thought. Therefore, it had been decided to allow for this when planning and designing new roads. What this meant in practice was that if a particular traffic forecast suggested a three-lane motorway under the old design criteria, under the new criteria a two-lane scheme might suffice. Alternatively, a single-carriageway road might replace a design for a dual carriageway. Thus, with a given sum of money buying more miles of improved road, the target mileage was repackaged. The aim now was 3100 miles of high-quality road by the early 1980s. Priority was to be "given to those road schemes . . . needed to complete a network . . . designed to meet the needs of heavy lorries".[15]

But a stable situation had still not been reached. With general economic circumstances deteriorating, further cuts were in the pipeline. By February 1976 an actual decline in trunk-road expenditure was proposed for the first time and, with it, the 3100-mile target receded to a new horizon – the mid eighties.

Now there were other problems with the Government's lorry strategy. The Department had carried out studies which showed that diversion to the network of 3100 miles would have increased lorry mileage to a significant extent, and the resulting costs were regarded as prohibitive. Therefore, in order to reduce the extent of the lorry diversion, the Department of Transport proposed in January 1976 a larger network of lorry routes totalling some 8000 miles.

This enlarged network included unimproved, unsuitable routes. Consequently, the new proposals were strongly opposed by environmental groups. In fact these groups were by this stage shifting their position. Although instrumental in the original idea of a lorry network, some were beginning to see clearly the full implications of this suggestion. As the Civic Trust was to note:

> "The routes proposed as lorry routes already carry heavy traffic and designation as a lorry route network would have increased the burden. But to by-pass all the settlements on the [8000-mile] network would require a road building programme of an unprecedented nature."[16]

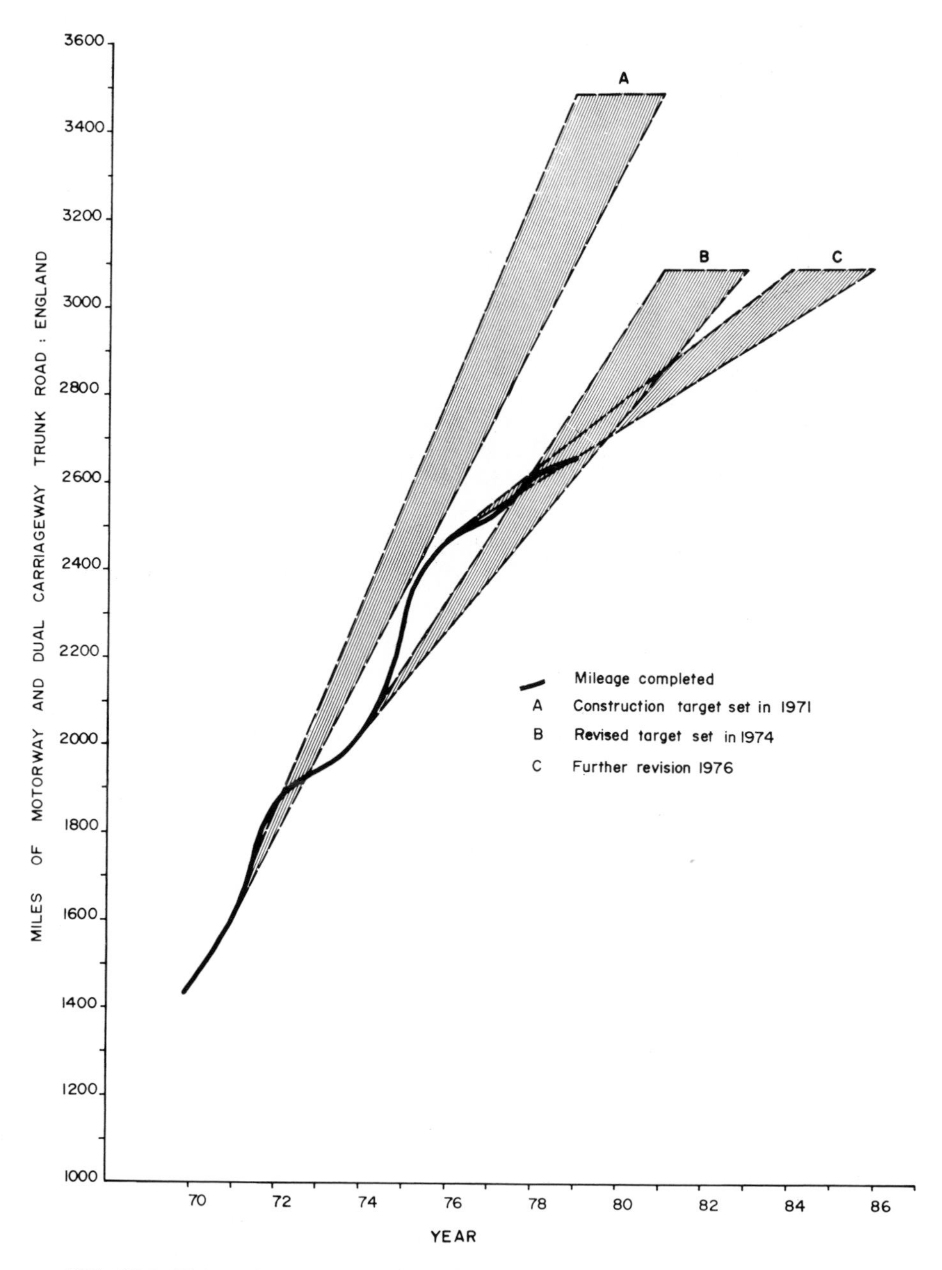

FIG. 12.1. High-quality trunk roads, targets and performance: as actual construction slumped, target mileage was revised on two occasions after 1971 and then finally abandoned in 1977.

Consequently, the Trust, together with other environmental bodies, eventually opposed the basic idea of a mandatory system of lorry routes.*

While the Department negotiated with local authorities and environmental groups, the Dykes deadline was fast approaching. Indeed, by this time many counties were in a difficult position. Those which approached their responsibilities under the Dykes Act with enthusiasm faced the difficulty of trying to co-ordinate their own network with a far-from-determined nationwide scheme. The unenthusiastic counties, however, did little more than aim to publish the one draft order required by 1 January 1977, leaving the matter of a strategic approach in the hands (for the time being at least) of Whitehall. This exacerbated the potentially difficult problem already beginning to surface of co-ordination between adjacent counties.

Oxfordshire for example disagreed with Berkshire over the use of the A4074. Berkshire proposed restrictions on the A417 to prohibit Oxford–Reading lorry traffic from using the poorly-aligned Thames Valley route running through sensitive settlements such as Pangbourne, Goring and Wallingford. Berkshire's preference was for heavy lorries to use the M4 and A34. However, a more direct alternative was provided by the A4074 and A423 mainly through Oxfordshire, but difficulties at the Reading end of the route led to Berkshire favouring bans on this route too. This was opposed by Oxfordshire.

Berkshire had its differences also with Surrey over proposals for a cordon of restrictions around Windsor. These plans proposed total prohibition on heavy vehicles crossing Windsor Great Park, but this would have led to greater use of Surrey's roads. The lorry lobbies were also strongly opposed to this scheme, and when it finally went ahead (the final order was made under the 1967 Road Traffic Act and not under the Dykes Act) they challenged, unsuccessfully, its validity in the law courts.

January 1st 1977 came and passed without any firm decision having been reached by the government on the nationwide proposals. Three weeks into the new year, William Rodgers, the Secretary of State for Transport, admitted to the House of Commons that there was less enthusiasm among local authorities and environmental groups for the idea of a national system of lorry routes than might have been expected: but still no firm decision. The position was repeated in March, when John Horam, one of the original supporters of Dykes' Bill and now Under Secretary of State for Transport, told the House of Commons that the Government had not yet decided whether to proceed with the proposals for a national network of lorry routes. By now of course it was nearly three years since the release of the original Consultation Paper, *Routes for Heavy Lorries*.

*The Local Government Operational Research Unit noted the ironic situation that some organizations which were traditionally anti-lorry in attitude opposed lorry routeing.[17] These were principally organizations aiming to promote other means of carrying freight, and presumably they felt that any scheme making lorries more acceptable would weaken their case.

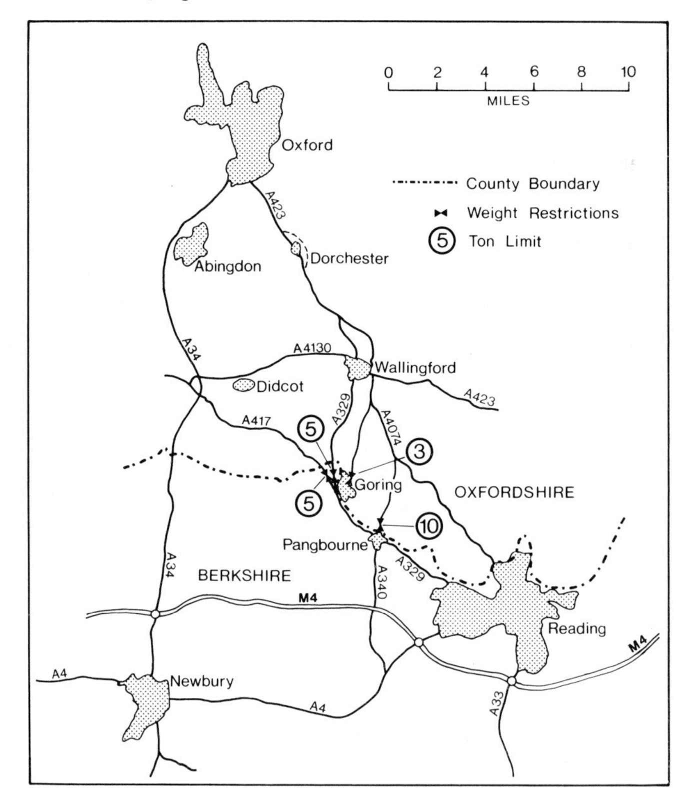

FIG. 12.2. The Goring Gap lorry control scheme: Berkshire County Council imposed two weight restrictions on the A329 and A417 near Goring. These were intended to encourage heavy lorries to use the M4–A34 route between Reading and Oxford. However, the A423–A4074 route provided a legal alternative on which Oxfordshire County Council refused to impose restrictions. This alternative route caused environmental problems in north Reading. Construction of a bypass on the A423 to remove heavy traffic from the centre of Dorchester began in mid-1981. The 10-ton limit is on a private toll bridge crossing the Thames.

Finally, in mid year a decision was made. The decision formed part of a more fundamental reappraisal of the Government's approach to trunk-road construction, contained in the June 1977 White Paper on Transport Policy. The White Paper noted that "the concept which has in England dominated a decade of road building has been the 'strategic network', a system of roads of high, uniform quality linking all the major towns and ports".[18] (It was, of course, a concept which in theory at least translated easily into the related concept of a national network of lorry routes.) Now this strategic approach was to be modified. It was no longer "a matter . . . of building to lines superimposed on maps and to rigid standards but of deciding on the right standard of each section of route". The keynote henceforth was selectivity, an incremental approach taking the most demanding schemes first and improving the network bit-by-bit. Consequently, the concept of a national network of lorry routes was abandoned, or, as the White Paper euphemistically put the point:

> "The Government has concluded that a national system of roads specifically designated as lorry routes will not be a practical proposition for some time."

The county councils were left to pick up the pieces. At the beginning of 1977 the Freight Transport Association reported[19] that many local authorities had countywide plans for lorry routes, either to the extent of having published a definitive map, or having prepared provisional schemes pending the Government's decision on a national network. Now with the national network gone it was that much harder for the counties to sustain their proposals in the face of often stiff oposition from lorry operators and, sometimes, the police. A few, like Berkshire and West Glamorgan in Wales, persisted with large-scale countywide measures. But generally speaking, the picture henceforth was of a local, *ad hoc* approach – in fact not too dissimilar in principle from the Government's new policy on trunk-road development.

Summary and Conclusions

Although the idea of restricting lorries to particular routes predated the juggernaut controversy it was the latter that provided the real stimulus for action. The environmental lobbies responded to the growing lorry-menace in the early seventies with suggestions for designated routes, and seized the initiative by promoting Dykes' proposals in the House of Commons. As conceived in its original form, Dykes' Bill proposed a National Advisory Council to stitch together local proposals on routes and restricted areas into a coherent nationwide system.

The Government's basic response, and one that was to set the pattern for the rest of the decade, was to argue that the problem could not be solved in the context of existing roads. Instead the real resolution was to build a suitable lorry network. The point was emphasized on many occasions and undoubtedly it helped to maintain the momentum of the trunk-road programme at a time when its urban counterpart was stalling.

Unfortunately for the Government's lorry policy, a combination of factors mainly of an economic nature but including also anti-road pressures from original proponents of the lorry-routeing idea, forced the inter-urban roads programme too into low gear. Eventually, proposals to establish during the 1980s a lorry network based on a specified improved network of primary routes was abandoned. Circumstances had thus killed what was thought to be the easier option of building the problem away.

Also abandoned, in effect, was a centrally determined lorry policy. Bypass plans to protect towns, villages and historic centres still existed, and with the shift in trunk roads policy in the late seventies these plans were re-emphasized. But routeing policy was now decentralized to the county councils: if there were suitable alternative routes for lorries, then local authorities were to make use of their traffic regulation powers to control heavy-lorry traffic.

However, this local approach was not without its problems. Potential routes crossing administrative boundaries were not always considered in equal light by adjacent counties. This was a problem made worse by the lack of enthusiasm for such ideas on the part of some councils. Indeed, the real significance of the Dykes Act was not that it confirmed new powers but that it forced even reluctant authorities to think about the situation.

However, even enthusiastic counties did not escape criticism from other quarters. The lorry lobby of course was strongly opposed to restrictions. But implied criticism of some local schemes came also from Sir Arthur Armitage in his Report on *Lorries, People and the Environment.*[20] Armitage noted his pessimism about the likelihood of getting net environmental benefits "at reasonable cost" from many large-scale schemes. In fact, he was to emphasize on a number of occasions the need to keep diversion costs "within bounds" and to avoid "unreasonable" detours.

Here, of course, lay the heart of the matter: the trade-off between hauliers' costs and environmental benefits. Different parties naturally had their own view of what was reasonable and, like Armitage, they did not always make this explicit. Therefore, it was unfortunate that in the mid seventies one opportunity to raise the level of the debate was not seized upon.

When the Department decided that the 3100-mile English lorry network proposed in 1974 would impose prohibitive costs on the haulage industry, its analysis showed that these costs would have increased by £130 million per year at then current prices (less than 1·5 per cent of the then annual nationwide expenditure on all road freight transport). At that time heavy lorries were failing to cover their contribution to the costs of providing and maintaining the road system to the extent of £110 million. The channelling of heavy vehicles onto more-suitably constructed roads would probably have reduced this cost but would have incurred additional expenses to administer and police.

As far as is known, these pluses and minuses, together with the number of households gaining and losing, were never brought together in the fashion of a balance sheet. Had they been, then the net costs might not have been looked upon as quite so prohibitive, and the idea of a nationwide network of lorry routes might not have disappeared quite when it did.

CHAPTER 13

"The Last Motorway?"

THE very month the Dykes Act reached the statute book, an auspicious public inquiry into proposals for the M42 motorway on the outskirts of Birmingham was drawing to a close. The inquiry was significant for two reasons. First, the M42 was the last motorway traceable on the tea room plan of 1946 still to be translated into hard bitumen. Second, at this eleventh hour, objectors raised as a major issue their rights to object on the grounds that a motorway was not needed. It was the start, in fact, of a long-protracted legal battle stretching over many years and culminating in what became known as the "Bushell case". At the one extreme were persons such as John Tyme, member of the Conservation Society. Tyme believed that the motorway and trunk-road programme with all its ramifications posed "a consummate evil" and constituted "the greatest threat to the interests of this nation in all its history".[1] At the other extreme was the Department, struggling unsuccessfully in the face of disruptions, walk-outs and lengthy adjournments to its various inquiries, to keep the roads programme on course and on time.

Blot on the Landscape

Opposition to rural schemes was not unknown in the early days of the post-war trunk-road programme. In the late fifties there was a short delay on this account in fixing the line for the M1 in Leicestershire, and in the early sixties there was the longer, celebrated delay in determining a route for the M4 between East Berkshire and Gloucestershire. The search for the latter had begun in the spring of 1961, but each line proposed – there were three main contenders – encountered considerable opposition from various amenity societies. It was indeed a difficult region in which to determine an acceptable path, with the Berkshire Downs, Vale of Whitehorse, Kennet Valley and Goring Gap all lying astride at least one of the proposed routes. Nevertheless, the case for a motorway along the broad axis proposed, was at no stage challenged. The thrust of the opposition was always to try to shift the proposal into someone else's back-yard; not to try to deter entirely the environmental impact of the motorway by questioning its need, but to seek to nullify its potential excesses by modifying its route.[2]

In the early 1970s this benevolent attitude was to change. The environmental backlash against motorways was by this time spilling over from the towns into the countryside and there were increasing calls for a termination of the rural motorway programme and for a review of transport policy. In late 1971 a coalition of environmental groups, called the Transport Reform Group, was formed to lobby for such a review. Its initial members were the Council for the Protection of Rural England (who had opposed particular routes of, but not the need for, an M4) and the Civic Trust. One of its early initiatives was to create the Midland Motorway Action Committee to co-ordinate opposition to motorway proposals criss-crossing the Midland region, and it was this Committee which first raised, in the context of the M42, the question of "need" for particular motorway schemes.

The Government's reaction to this initial challenge was confused and ambiguous, a situation that was to prevail for some time. Prior to the inquiry, the M42 objectors had been led to believe by the Parliamentary Under Secretary of State at the Department of the Environment that the question of "need" could

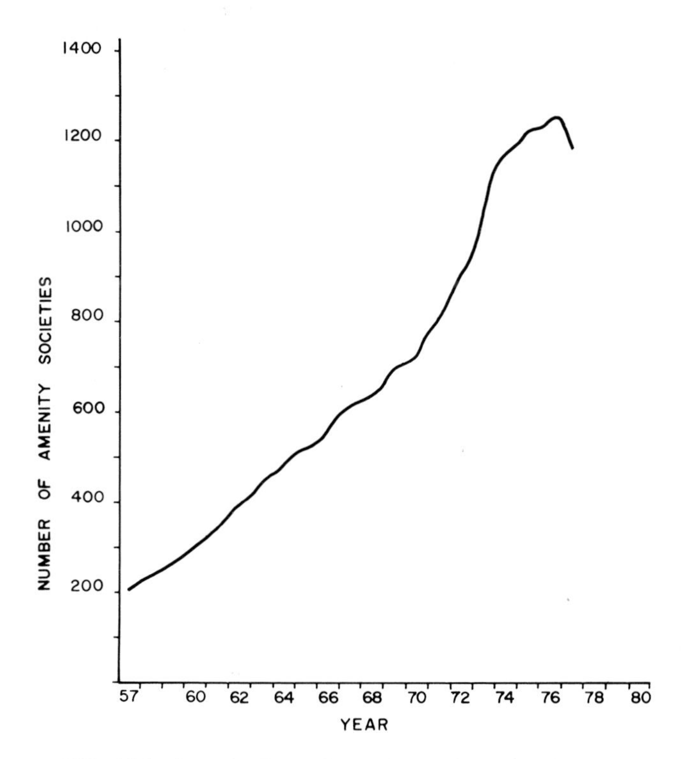

FIG. 13.1. Growth of amenity societies 1957–77: the illustration is based on the number of societies affiliated to the Civic Trust. The period of most rapid growth came in the early 1970s.

be raised, only to be told subsequently that the issue was the "need for the proposed M42 on the present draft line".[3] This position was further qualified at the inquiry when counsel for the Department argued that the decision as to what evidence and submissions were relevant was entirely one for the inquiry inspector. As it transpired, the inspector at the M42 inquiry refused to allow cross-examination of the Department's witnesses on the subject of the traffic forecasts which formed the basis of the scheme.

The issue was taken up the following year by the Select Committee on Expenditure, this time with a new Government in power at Westminster. The Committee came down strongly in the objectors' favour, arguing that local circumstances frequently merited a broad approach. They therefore recommended:

> "that the right to challenge the need for a transport scheme at a public inquiry should be firmly established. It should allow consideration of different pricing structures and of the traffic forecasts which underline proposals. It should not, however, embrace the question of directing passengers or freight on a particular form of transport; this is a matter of policy, to be decided by Parliament."[4]

The point made was clear but the Government reply was inconclusive if not ambiguous:

> "The Government agrees that the need for a road scheme may appropriately be challenged at a public inquiry, provided that matters of policy are not called into question. The detailed application of this principle is being studied."[5]

Any hopes for a considered change of attitude that the objectors may have had from this reply were quickly dashed. The following month, a revised *Notes for the Guidance of Panel Inspectors* was issued to members of the panel from which inquiry inspectors were drawn.[6] The gist of the note was to repeat the view that local inquiries were not appropriate venues for discussing matters of Government policy. It went on to argue that, in effect, publication of a scheme indicated that it was the Minister's policy to build such a road (though not necessarily on the draft line of the route). Consequently, it suggested that inspectors should protect Department officials from cross-examination on matters such as traffic forecasts used to justify schemes. These too were *de facto* policies.

Objectors reacted angrily to the manifestation of this view – the actual Notes were confidential and not released – and subsequently public inquiries began to get out of hand. In 1975 a series of major inquiries were seriously disrupted. This culminated in February 1976 in the indefinite adjournment of the inquiry into the proposed Airedale trunk road. By now the average time taken between the first preparation stages and the start of construction on a trunk-road scheme had slipped from five to seven years in the 1960s to between ten and twelve years by the mid seventies: the Government was alarmed by the delay.

The Inquisition

In order to defuse the situation, the Government announced in 1976 a number of developments. The most important of these was a proposal first mentioned in the Consultation Document on Transport Policy, namely to seek an independent review of the Department's methods of scheme appraisal, including its traffic forecasts and the relative importance it attached to economic and environmental factors. The proposal was acted upon later that year when, in December, the Advisory Committee on Trunk Road Assessment was formally announced. Its Chairman was to be Sir George Leitch, a former defence bureaucrat, and its other seven members were drawn from a wide number of sources including British Railways and the Civic Trust. The Committee's terms of reference required a review of the Department's trunk-road assessment methods, paying particular attention to the Government's traffic-forecasting methods and its economic analyses of road schemes.

At this time the Department's traffic forecasts were still based on work undertaken in the early 1960s at the Transport and Road Research Laboratory by a mathematician, John Tanner. It was his forecast of private-car traffic that was of fundamental importance. The essence of Tanner's approach was to assume that the number of cars per head of population would eventually reach an absolute level known as a "saturation level", determined largely by observing what had happened in the most affluent parts of the United States. He then postulated that car ownership and, therefore, car traffic – traffic was assumed to have a constant relationship with ownership – would grow from then existing levels at a rate depending upon how near to being reached was this saturation level. Once levels of ownership were high and, therefore, close to saturation level, the rate of growth in traffic would be low, and vice versa.

Tanner's original forecast was published in 1962, and it had a bearing on the Buchanan team's deliberations.[7] It was a fifty-year forecast that used 1960 as a base and it was greatly influenced by Government estimates of future population growth that proved, subsequently, to be wildly optimistic. Over the years, the original forecast was revised several times, although the mechanics of the approach used remained pretty much the same. The more detailed changes of method enabled estimates of future income growth and the real costs of motoring to affect moderately the rate of growth in car ownership, although not the saturation level. The latter was changed a little in 1965 but thereafter remained unaltered. Income estimates together with forecasts of future motorway mileage (the latter reflecting a higher quality of travel) were allowed also to influence, in a marginal way, vehicle use as opposed to ownership.

The Department's economic-evaluation methods, however, had changed character more frequently than its traffic-forecasting methods, but here too the roots were to be found in the early sixties. A great influence had been the London–Birmingham motorway cost benefit analysis (see p. 7). From this analysis the Ministry of Transport tried to develop a method that could be

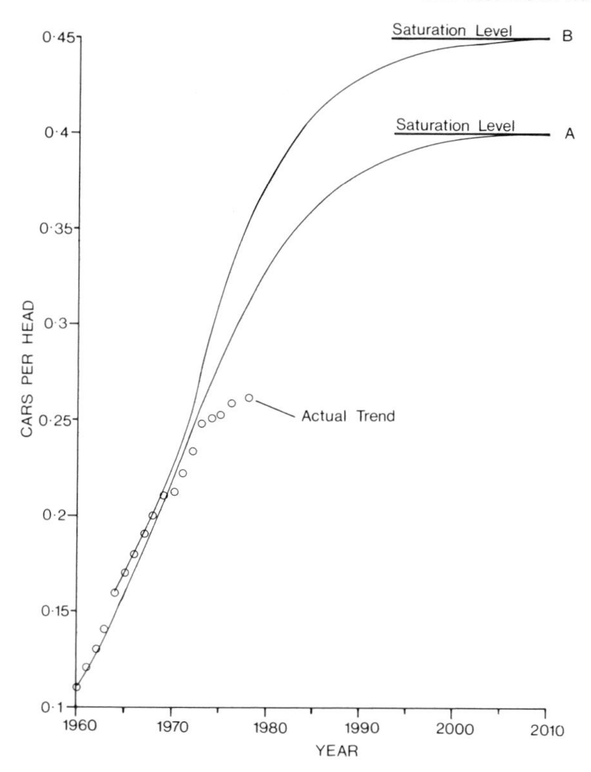

FIG. 13.2. Tanner's forecasts of cars per head: shown are the original 1962 forecast (A) and the first revision in 1965 (B). These car per head forecasts were multiplied by the corresponding population forecasts to give total car population at various dates.

applied much more easily to a large number of schemes. The initial outcome in the mid sixties was an approach which calculated losses of travel time and increased accidents (both expressed in money terms) resulting from a failure to improve an existing route to a higher standard.

Around 1967 this approach was superseded by ideas for comparing the benefits during the first year a proposed new road was likely to open, with its capital cost. In turn, this method was superseded in 1973 by a computerized routine known by the acronym COBA (cost benefit analysis) which allowed for easier application of a method that came closer to the original M1 study in terms of the components of cost and benefits taken into account. There was, however, one major difference from the 1960 study. An attempt was now made to take account of costs and benefits over a period of thirty years after the proposed opening date of the scheme, which, of course, gave added importance to getting accurate the forecasts of future traffic flows. Also, like the M1 study, it raised the vexed issue of appropriate values to attach to the various items of benefit, the value of time saved on a journey being the most difficult to assess.

Leitch reported to the Secretary of State for Transport, William Rodgers, in October 1977, within nine months of starting work.[8] But such speed did not detract from what emerged as a substantial and considered analysis of a complex issue. With regard to existing methods of traffic forecasting a number of points were made. In conducted tests, the "saturation level" appeared to have a strong influence on the economic rate of return from road schemes. Yet the Committee found the lack of a firm theoretical basis for the chosen level disturbing. In addition, the Committee was not convinced of the correctness of the assumed path of growth towards the level of saturation. Moreover, an examination of observed, as opposed to forecasted, traffic flows on a number of schemes that had been open for some time, "seem[ed] to support the general conclusion that there had been a tendency in the past to over-predict traffic, in certain cases significantly".

They therefore recommended that an alternative approach be adopted by the Department, taking account more directly of the commonsense factors influencing car ownership and acknowledging the uncertainties inherent in such forecasts. Complementing this was a further suggestion that there should be more research into the factors affecting the use of cars. In this context they agreed with those critics of the Department that to use the length of proposed motorway in existing formulae to indicate the tendency for better quality roads to generate more traffic contained an element of self-fulfilment (although it was thought not of great quantitative importance). The Committee also suggested that regular and consistent use of studies to check the accuracy of past forecasts could prove most useful.

The cost benefit methods were considered sound as far as they went. However, the fact that the environmental effects of trunk-road schemes could not be taken into account in money terms led, they felt, to the assessment being dominated by those that could be so valued. Therefore, they wanted to see a shift of emphasis to an approach set in a broader framework where effects were systematically recorded in a general balance sheet of pluses and minuses, only some of which would be in money terms. They added that, as an approach, it emphasized judgement and the form of presentation was crucially important to ensuring that the issues were clearly set out. Clarity they considered of utmost importance if the Department's methods of appraisal were to command more public confidence than they had enjoyed in recent years.

The second major inquiry announced in 1976 (two weeks before the Government were forced to abandon the notorious Airedale inquiry in the face of disruption) was a review of the procedures surrounding the conduct of public inquiries into highway proposals. In contrast to the review of trunk-road assessment methods, this review was basically internal, undertaken jointly by the separate Departments of Transport and Environment, but in association with the independent Council on Tribunals. It resulted in a White Paper published in April 1978.[9]

The White Paper confirmed one major change that had been provisionally introduced pending the outcome of the review. This was the selection of inquiry inspectors by the Lord Chancellor rather than, as previously had been the case, their selection and appointment by the Secretaries of State for the Environment and Transport. The intention here was to assuage the suspicions that naturally occurred when the proponent of the scheme was also responsible for the appointment of the "jury".

The White Paper confirmed yet another major change that had been introduced after the "Review of Highway Inquiry Procedures" had been announced. This had been the introduction in the revised "Highway (Inquiries Procedure) Rules 1976", operative from June 1976, of the entirely new concept of "statutory objector". Statutory objectors were basically persons having a direct property interest in the proposed scheme (either because of land required or because of their entitlement to compensation under the 1973 Land Compensation Act). The idea was that only statutory objectors would have an entitlement to appear at an inquiry and give evidence. All other objectors could appear only at the discretion of the inspector. The Report of the Review upheld this concept on the grounds that it would help to avoid repetition or delay "by filibustering or disruption".

A material innovation suggested by the Review was to hold a pre-inquiry procedural meeting, the purpose of which was to agree upon a basic programme for the inquiry and to establish as far as possible policy matters considered relevant. In this context there had again been developments during the course of the Review, with the Government conceding the possibility of traffic forecasts being questioned on the basis of particular local factors. In other words, the Government was seeking to develop a distinction between, on the one hand, the fundamentals of its assessments which were determined at the national level and therefore were considered matters of policy, and, on the other hand, the interpretation of these fundamentals in a local setting. This was expressed in the Review Report as follows:

> "While it would be inappropriate for national policies, including technical methods which are determined nationally, to be re-examined at each local inquiry, the Department will explain why any forecasts and standards which have been applied to a particular road scheme are appropriate to that scheme in its local setting and it will be proper to consider at the inquiry whether they accurately reflect local circumstances. Opportunities will be provided for objectors to question the Department on these matters: similarly, the Department's representatives may wish to question objectors to test the validity of their arguments."

In order to facilitate this approach, there were also proposals for objectors to have considerably increased access to information. The information to be provided was related to alternative proposals in addition to the route preferred by the Department and it was to include ". . . the facts and assumptions on which the case for the scheme is based; results of analyses; explanations of methodologies; and any other information which . . . can be provided without costly special research".

The Consequences

Throughout the long controversy, an important plank in the Government's case had been that its assumption about the future availability and price of fuel, future car ownership and the priority given to roads relative to other transport modes, were issues of policy more appropriately debated by Parliament. This had brought forth the response at a number of inquiries – it was in fact to form the substance of John Tyme's objections – that such issues had not been put before Parliament and hence subject to debate.

The 1976 Consultation Document on Transport Policy proposed, therefore, an annual White Paper which would set down in clear terms the policy upon which the roads programme was based. The first of these White Papers was published in April 1978 (concurrent with the White Paper on the Review of Highway Inquiry Procedures) and the second in June 1980.[10,11] Between them they represented a considered response to the reappraisal of methods and procedures, and, in the light of the new approach adopted, a review also of individual schemes in the roads programme.

Both White Papers accepted in principle all of the Leitch Committee's major recommendations. On the subject of traffic forecasts, new interim forecasts were made available in January 1978 when the Leitch Report was published (it had been received by the Secretary of State the previous October) and a revised forecast was published in July 1980. As suggested by the Leitch Committee, these new forecasts did not attempt to predict precisely what would happen in the future, but for the first time recognized the inevitably uncertain nature of forecasts by using a range of assumptions giving, in turn, a range of estimates of traffic growth. With regard to the Advisory Committee's suggestions on a new framework for appraisal, the Department introduced a number of experimental frameworks which were tried out in relation to a number of schemes. As a result of these trials, a specific, favoured framework emerged in 1980 and was gradually applied as schemes came forward in the roads programme.

The 1980 White Paper, like its 1978 predecessor, also confirmed the Government's acceptance of the recommendations of the Review of Highway Inquiry Procedures. By now the Government's policy on procedure had withstood challenge in the courts,* and consequently the White Paper repeated the point

*Objectors to the M42 motorway had appealed against the Minister's decision not to re-open the public inquiry on two grounds: first, the refusal to allow cross-examination of

made earlier that it was not considered sensible for the basis of Government roads policy to be examined at each of the thirty or more trunk-road inquiries held each year. But it added that it was appropriate for inquiries to question whether local factors might result in traffic growing faster or slower than the national average, or whether the relief provided for existing roads would be more or less than the level assumed. It added, "This is the element of 'need' for a road scheme which the main inquiry into any scheme can sensibly consider".

As a result of this general reappraisal of the roads programme, the Department set out to review all schemes in the roads programme. There were two main aspects of the review. First, to reconsider the need for and standards of the scheme (particularly in the light of the revised traffic forecasts), and second, to consider the actual timing of schemes in the programme (bearing in mind the different emphasis now given to environmental aspects in the appraisals).

There were, as a consequence, a number of changes. The 1978 White Paper, for example, noted that thirty-four schemes costing in all about £90 million were to be replaced in most cases by bypasses and small improvement schemes at a much lower cost. Specific examples were the M54 to Telford new town which was to be designed as a two- rather than three-lane motorway, and a link between the M6 and Blackburn in Lancashire which was no longer to be a new highway but a scheme improving the existing road. The 1980 White Paper extended this picture by drawing attention to a number of large schemes that had been deferred in order to give priority to a number of bypasses that would remove heavy-lorry traffic from towns and villages. Of the deferred schemes the largest was the Maidstone–Ashford section of the M20 motorway.

The general picture was one of slightly less emphasis on motorway schemes and a steady shift towards bypasses for historic towns and for market towns, a trend that was expected to gather pace during the eighties. Indeed, as the Parliamentary Secretary at the Department of Transport remarked early in 1980:

> "obviously we are not far away from the end of the motorway building programme. The inter-urban network . . . is not yet complete, but there will come a stage, and it is probably a decade or so away, where those routes that justify links of motorway standard will be more or less complete."[12]

A few months later he was to prophesy that the end of the programme would coincide with the completion of the M40 extension from Oxford through to Birmingham.

An optimistic note was sounded too by the Deputy Secretary at the Department of Transport responsible for the roads programme. In this instance the

---

the Department on its traffic forecasts was contrary to natural justice; second, that there had been important developments since the inquiry closed. The appealant, John Bushell, lost his case in the High Court in 1977 but triumphed in the Court of Appeal in 1979. The House of Lords overturned the Court of Appeal judgement in 1980.

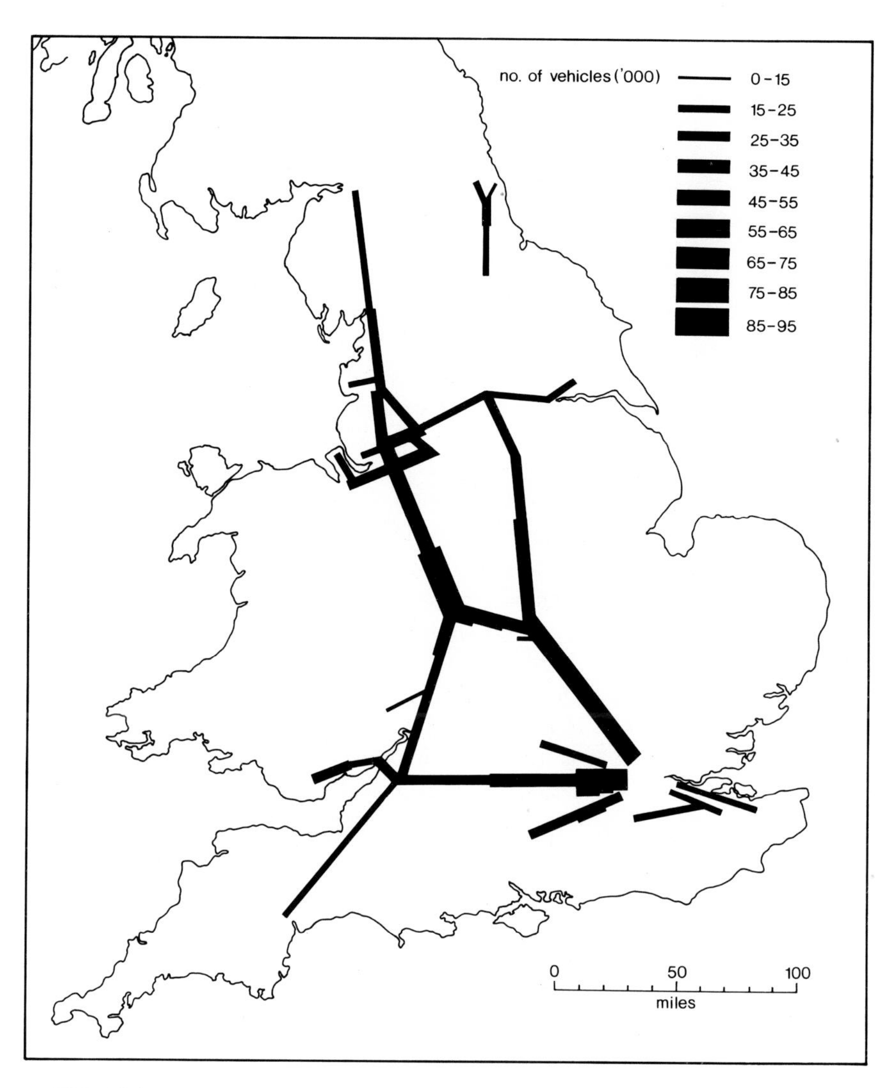

FIG. 13.3. Traffic flows on selected motorways, 1978: shown are the daily number of vehicles. The least used motorways were the M45 and M50. Parts of the M4, M6 and the M1 were the busiest.

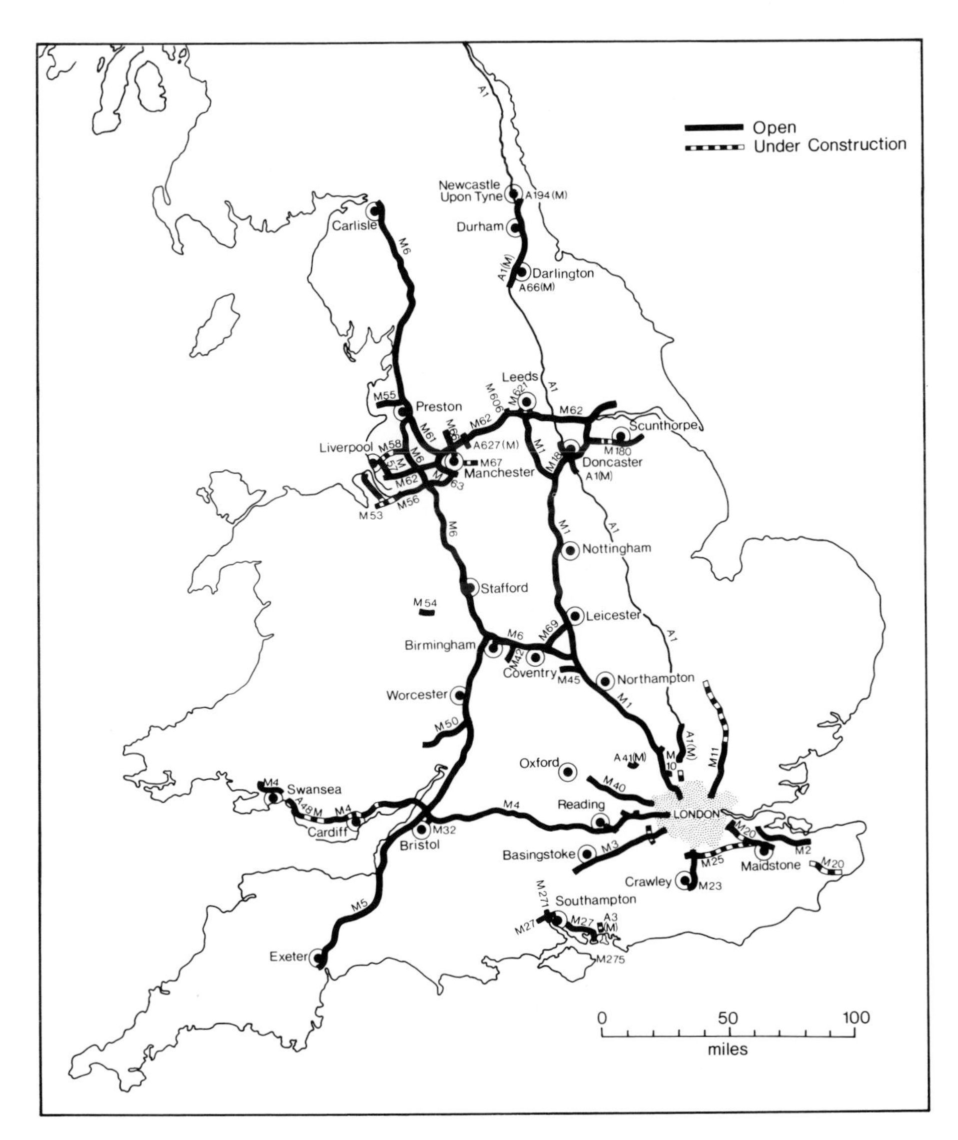

FIG. 13.4. Motorways in England and Wales, December 1978: there remained a strong correlation with the 1946 plan of the post-war Labour Government.

optimism related to road inquiries and the Deputy Secretary's detection of an apparent change of mood on the part of objectors. He noted that under the new inquiry procedures there had been only limited disturbance at the first inquiries and "almost none at later inquiries".[13]

But the omens were not all so favourable. The environmental lobbies appeared, on the basis of their evidence submitted to the new House of Commons Select Committee on Transport (the consequence of a Parliamentary committee system radically altered in 1979), to be aggrieved still on a number of points. For example, it was thought most unsatisfactory that officials of the Department's regional Road Construction Units (these were established in the sixties in order to streamline the road design bureaucracy) might be called upon to give subsequent advice to the Minister on the inspector's report. Moreover, in spite of the Leitch Committee's strictures on the importance of clear communication of issues if the Department's approach was to command respect, the Select Committee was confused still and referred to the Department's process of making decisions on road schemes as "a highly arcane art masquerading as science". They added:

> "even those professionally involved in questions of transport policy appeared to find it hard to understand how national policy for road building . . . is formulated and how overall national policies are in practice translated into decisions to build, or when to build, individual roads."[14]

In these circumstances, the Parliamentary Secretary's optimistic prognosis on the timing and particulars of the last motorway may have proved to have been a little premature.

Summary and Conclusions

The antipathy towards rural motorways that evolved during the seventies revolved around a number of related issues. One such issue was a deep-rooted scepticism shared by most of the environmental lobby towards the methods that the trunk-road planners employed to justify their road schemes. The traffic forecasts for example were thought to grossly exaggerate the situation likely to prevail. There was also an increasing disenchantment with the economic analysis, reflecting in part a wider adverse reaction to cost benefit analysis in the wake of the contentious Report by the Roskill Commission on the siting of the Third London Airport. The manifestation of these growing doubts was an attempt to question the "need" for particular road schemes. Once the Government had rejected this challenge, largely on constitutional grounds, it was only a matter of time before matters came to a head with the disruption of road inquiries.

The Government sought to diffuse the issue by announcing in 1976 an independent review of its road appraisal methods and a review (rather less independent) of the procedures for holding inquiries. It also promised a regular White Paper setting out the *raison d'être* of the roads programme. In 1977 the first review emerged – the Leitch Committee's assessment of the Department's techniques and assumptions. The Committee found the Department's traffic-forecasting procedures unsatisfactory and its economic appraisal of schemes too narrow in focus: a judgement which was promptly hailed by several of the objector groups as a vindication of their views. The thrust of the conclusions in the Report of the second review – the Review of Highway Inquiry Procedures – was on the one hand, to try to increase the apparent fairness of the proceedings, but on the other hand, to maintain a distinction between such matters as national car ownership forecasts and the interpretation of these in a local context. The former, the national forecasts and assumptions, were to be excluded from consideration at inquiries, whilst examination of the local implications was permissible.

The first White Paper on Roads Policy emerged in 1978 and the second after a change of Government in 1980. Between them they portrayed a roads programme adjusted in the light of this general review, of the comments received and of the subsequent developments. The resulting picture was one of less emphasis on a network of long-distance motorways and of more emphasis on small-scale bypasses. Nevertheless, motorway schemes were still very evident and, from the pattern of those still outstanding, it was quite apparent that one major conclusion of the Leitch Committee had not been accepted.

The Leitch Committee had grappled with the very fundamental issue of whether there could be extra benefits, not reflected in the predicted flows of traffic and hence not captured in the normal cost benefit procedures, as a result of induced regional development. They had come to the conclusion that there were not. Such a conclusion, of course, ran counter to the ethos which had initiated the post-war trunk-road programme. There had been in fact many examples of large-scale road schemes unjustified by predicted traffic flows which, nevertheless, entered the roads programme in the cause of furthering regional development. The Ross Spur – A40 – Heads of the Valley improvements were decided upon in order to try to rehabilitate the Swansea area of West Wales; the Scottish M8 failed the normal criteria and was justified on presumed development grounds. Similar considerations related to the M90 between the Forth Bridge and Perth, and also to the reconstruction of the A9 Perth-to-Inverness road, "the lifeline of the Highlands".[15] In England, comparable examples came from the North East where a special roads grant had been decided upon in 1962:

> "not so much because of existing traffic problems, which in fact are in general less severe here than in many areas elsewhere, but because of the contribution which a forward looking road plan can make to the region's faster growth."[16]

And of course there was the tale of the two bridges across the Forth and Humber estuaries. Now it seemed, notwithstanding the Leitch Committee's view, the story was to be repeated in the eighties. This time the schemes in question included the North Devon link road and the M54 motorway to Telford. In this latter case, following a public inquiry in February 1976 on a proposed section of the M54, the inspector concluded that he could see no urgency for building such a scheme and recommended rejection. However, the Secretaries of State (Transport and Environment) subsequently put aside the inspector's advice.

But really such decisions merely underlined a basic shortcoming in the scope of the Leitch Committee's review. It failed to address (and its terms of reference were responsible in part) the crucial issue of the role of analysis in the final process of making decisions. There had long been a strong presumption (which the Leitch Committee was to share) that, first, analyses of traffic flows and, then, a more sophisticated approach using economies, played an early role in developing the trunk-roads programme. It was a view that the trunk-road planners fostered. It had brought approval from the Treasury and it had helped the Government to maintain the view, put to the House of Commons Expenditure Commitee in 1972, that the cost of the inter-urban road programme was fully justified by the returns obtained from individual improvements.

However, the realities of the case were rather different. For a start the combination of long gestation periods for highway schemes and the time taken to develop a standard evaluation procedure meant that economics had had little influence on the shape or sequencing of the trunk-road network *constructed* by the early seventies. As the House of Commons Estimates Committee had been reminded in 1969, road schemes then starting had been firmly programmed (five years before) without reference to economic criteria. And, by the time that COBA had been established and could begin to influence road planning, over 2000 miles of motorway and other high-quality trunk road were already complete.

But it is a moot point also whether COBA, once fully established, had much of an impact on the shape and size of the recent roads programme. The decision to invest in a road scheme was at the end of the day a matter for judgement alone. The technical analyses played a part in that decision, but factors which techniques of measurement did not cover were taken into account also. How and to what extent they were taken into account was a matter left unclear. It was a point that the Leitch Committee did not (or could not) address and it remains unanswered today: a prerogative of politicians – or mandarins?

CHAPTER 14

Perspectives on the Policy Process

THE first point to note about roads and traffic in post-war Britain is that, for the most part, governments had a positive policy towards them. That is to say, there was a basic idea of where policy was leading and what was being aimed for. Sometimes these aims were expressed formally, for example, in the shape of a master plan for trunk roads. Sometimes they found expression in a specified course of action, such as encouraging local authorities to adopt traffic studies. And on other occasions there was an implied target such as maintaining, or improving upon, traffic speeds in Central London.

If all this seems self-evident then we need only reflect on Plowden's verdict on government attitudes towards the motor car.[1] In Plowden's opinion, for three-quarters of a century governments had failed to take any kind of comprehensive view. Official policy was, in his terms, a "non-policy". Indeed, it must be said that at times road policies also had exhibited similar characteristics. During the fifties, for example, the attitudes of governments towards urban roads and traffic (in marked contrast to policies for trunk roads) were distinguished by a complacent apathy. This disinterestedness was an expression of a belief (verging on hope) that traffic would find a satisfactory level without the active involvement of governments. It was, in Plowden's terms, a case of choices not made by governments, and consequently of decisions made by default. But on the whole, for most of the post-war period the roads picture was a positive one. Policies for the most part were imbued with a sense of direction.*

The second point to note is that the policies which have prevailed since the Second World War have been long-term policies. For example, the policy for towns and cities that emerged at the beginning of the sixties (once the apathy of the fifties had faded) was to last for more than a decade, the first distinct

*This sense of direction was not always evident to the outside observer, or, if it was evident, it was sometimes felt that it was not given strong enough expression. For example, where responsibility for implementing a policy rested with local government (and this was generally true outside the realm of trunk roads), then Whitehall's style was often to adopt a low profile and apply discreet pressure. The issue revolved around the conflicting concepts of strong central direction on the one hand and local democracy on the other. The 1972 House of Commons' Expenditure Committee's Report on Urban Transport Planning provided one example of central government being called upon to give firmer direction.

change coming in 1973/74. But quite exceptional in terms of longevity was the initial policy on trunk roads. This spanned at least two decades (from Lennox-Boyd's statement in 1953 to the first cuts in trunk-road spending in 1973). Indeed, taking a more generous view, it is arguable that it lasted from the launching of the tea room plan in 1946 through to the latter half of the seventies. In this latter period, a combination of cuts in the level of spending and a shift of emphasis from an integrated motorway network to individual bypass schemes could be said to have finally foreclosed the initial post-war policy.

The important point here is that these policies were not long-term by accident but by design. There was a long-term strategy guiding the day-to-day decisions. In the urban context, for example, local authorities during the sixties were encouraged to introduce what was then seen as a short-term expedient of traffic management; they were encouraged also to conduct large-scale comprehensive traffic surveys at a time when many councils were anxious to start building roads. But behind such initiatives was the long-term strategy of first completing the 1000-mile trunk-motorway network and then diverting more resources to urban roads.*

The third point is that policies did change over time. The manner of the change, however, was interesting. Policies evolved, but in a sense they evolved rather less than might have been expected. Change when it came was a change of substance, which resulted in a fairly distinct break with the past. In some instances, the distinction arose by a change to a positive policy (encouraging traffic management) from a "non-policy", or from a positive policy (a national network of lorry routes) to a rather negative, ill-defined policy. Alternatively, the change was from one positive policy (building urban-motorway networks or a strategic network of trunk routes) to another policy less precise but none the less positive (selective trunk-road improvements or managing the demand for urban road space). Thus significant changes in policy in the long-term occurred not as a result of a continuous process of small, incremental changes over a long period of time; on the contrary, the process of change was itself rather more positive and quite different from what has been purported to be the "British style" of bit-by-bit adjustment to a policy not always well defined.[4]

The Relevance of Parties

What brought about such distinct changes in policy? One thing is clear (and it

*It is interesting to note in this context that on the basis of recent statements, the new trunk-roads policy is *expected* to last for a further ten years. In addition to the Parliamentary Secretary's comments foreshadowing the end of the motorway programme in a decade or so,[2] the Government Observations on the Transport Select Committee's Report on the Roads Programme commented that "the Department has a major role to play in this [trunk roads system] for the next decade. Beyond that . . . it may be possible to reduce the scope of the trunk-road network and appropriate to review arrangements for administering it . . .".[3]

constitutes what is probably the most significant outcome of this review of post-war road and traffic policies): it was not brought about by a corresponding change in the party of government. In fact politics in the conventional sense of party politics appeared to play very little part in the formation of post-war road policies at the national level. All the major changes in the direction of policy outlined in the preceding chapters occurred towards the end of a particular administration, whilst the continuity of policy from one political administration to another was overwhelmingly strong. Thus, the first post-war Conservative Administration adopted the Attlee Government's tea room plan for trunk roads; the Wilson Administration adopted the urban road traffic policies honed during Marples' years at the Ministry. In 1971 the Conservatives adopted the revitalized trunk-road strategy fashioned in the dying days of the 1964–70 Wilson Administration. Similarly, the Labour Administration in 1974 adopted the recast urban policies moulded in the last two years of the Heath Administration. And, to advance the story still further, the fundamentals of the 1977/78 reappraisal of trunk-road policies were adopted by the current administration of Mrs. Thatcher.

In fact it is difficult to think of a single instance of anything that could be said to constitute a "reversal" of a previous administration's policy on party political grounds. It is true that at the beginning of the Heath Administration in 1970 different points of view appeared to emerge. One such example was the juggernaut lorry issue and Peyton's announcement of December 1970 distancing the new Government from pressures to increase lorry sizes. It is doubtful, however, that if the Labour Government had been in power a different decision would have been made. Worthy of note here is what Mr. Mulley had to say during the Commons debate in November 1972:

> "When I was Minister of Transport I was immensely impressed by the representation which I had from . . . [the opponents of larger lorries] . . . Consequently I caused a full examination to be initiated. I was not in office when the reports came back, but I like to think that they had some part in persuading the Minister for Transport Industries to take what everybody conceded at the time was the right decision on that recommendation."[5]

Of course, it is easy for politicians to say what they would have done, given the power and the benefit of hindsight, but the circumstantial evidence does point only to an apparent difference between the parties on this issue.

The two different parties did, of course, carry out policies in their own styles. Some policies were pursued with greater enthusiasm by one party as opposed to the other, but here again in the circumstances the differences were perhaps not as great as concluded by some observers. Thus, Braybrooke's comment[6] that the Labour Government in the sixties, in contrast to the Tories, pursued with vigour the road-pricing issue, is not really proved. One has to take account here of the different circumstances faced by each party when in government. At the time when the Smeed Report on road pricing reached Whitehall's desks, the Conservative Administration were aware that they were soon to defend their majority in a General Election. Therefore, it would have been inadvisable for them to have been seen to support the Smeed Committee's contentious conclusions at such a sensitive time. Whether they would have continued to react circumspectly had they won the election is mere speculation, but as Mrs. Castle pointed out in a press interview in 1967, "Don't forget that it was Ernie Marples who started all this road-pricing lark".[7]

At times, parties tried to give the impression that there was more of a difference between them than was really the case. Thus the 1976 Consultation Document tried to draw a marked distinction between Labour's approach to roads spending and that of the previous Conservative Administration:

> "The programme for England, included in the figures in the previous Administration's White Paper of December 1972 (Cmnd. 5178) required an increase in expenditure between 1971/72 and 1976/77 on trunk roads and motorways of 62%, on other roads 19% and on maintenance of 32%. For all these categories the Government are proposing decreases instead of increases for the next five-year period."[8]

This was true as far as it went, but the December 1972 White Paper on Public Expenditure was not the last such White Paper of the Heath Government. The last one was in December 1973 and that too showed a decrease in spending on principal and other roads.

Whether such attempts at political window-dressing subsequently made their mark with the electorate is an arguable point. But much as the national parties may have tried to steal from their opponents the "middle ground", behind the façade the measure of agreement between the two major parties was really quite considerable. When it came to elections and the publication of the manifestos, the consensus was then quite apparent.

In the October 1974 Election, for example, both parties appealed to the electorate with a policy of lorry routeing. In 1970 both parties promised an expanded inter-urban roads programme. In 1964 the Conservatives promised to apply the principles of the Buchanan Report, whilst the Labour Party promised urgent attention would be given to Buchanan's proposals and to the development of new roads capable of diverting through-traffic from town centres. In 1959 the Conservatives drew attention to their road programme, "the biggest we have ever had in this country", and promised even more; Labour, however, considered the Government's road programme "entirely inadequate" and also promised more.[9]

The Influence of Pressure Groups

There is now a fashionable theory that policies in Britain are influenced to a considerable extent, even formulated, as a consequence of the activities of pressure groups or lobbies. Included in this description are the senior civil servants – the mandarins – who have been referred to as constituting the biggest pressure group of all. This latter view was greatly stimulated by *The Diaries of a Cabinet Minister* wherein Richard Crossman expressed the view that "the Minister is not merely subject to control by his own Department which seeks to make him work according to their department policies", but also to a wider network of Whitehall control exerted through official committees.[10] More recently Tony Benn has fanned the same flames of controversy.[11] Specifically in the context of road and traffic policies, the issue surfaced in a rather spectacular way in 1978 with the leaking of what is now known as the "Peeler memorandum" – a note written by Joseph Peeler, an under secretary heading the Department of Transport's freight directorate on the subject of an inquiry into heavy lorries. The note canvassed the idea of an inquiry on the subject of higher weight limits for lorries as "a means of getting around the political obstacles". The note also made specific reference to hopes for "a conclusion in line with the department's view"[12] as distinct from the Minister's view.* In doing so it confirmed the preconceptions of many.

Whitehall mandarins clearly do have considerable powers either to make or to modify government policy. However, all those general factors listed by Peter Kellner and Lord Crowther-Hunt[14] which add to this potential were equally evident in the case of the post-war transport portfolio. Transport ministers, for example, came and went at intervals no less frequent than experienced in other sections of Whitehall. Of the fourteen terms of office between 1945 and 1979, six were for periods of eighteen months or less.† It is perhaps indicative that none of the six terms was associated with the major changes in the direction of policy outlined previously.

In contrast, Permanent Secretaries of the Department of Transport and its predecessor "ministries" saw longer terms of appointment. Only one since 1947, Sir Ian Bancroft, enjoyed less than three years, whilst Sir Gilmour Jenkins held the position from 1947 until 1959. However, the Jenkins case was exceptional, and three to four years appears to have been a common term during the last twenty years or so. Nevertheless, even these moderate terms of appointment were longer than most ministerial terms, giving the mandarins all those advantages associated with greater familiarity with a portfolio.

*A less spectacular but no less revealing episode occurred in 1966 when it was reported in *The Times* that Mrs. Castle, whilst in New York, favoured a road-pricing system for London based on direct charging for the use of roads. In a related news story under the heading "London denial of any firm decision" the motoring correspondent reported that the report from New York "caused surprise in official circles in London yesterday . . . I am reliably informed that no firm decision has yet been taken on the exact system of restraint that may be adopted in London".[13]

†Mr. Mulley, however, had two terms of about a year each, the last one as Minister for Transport within the Department of the Environment.

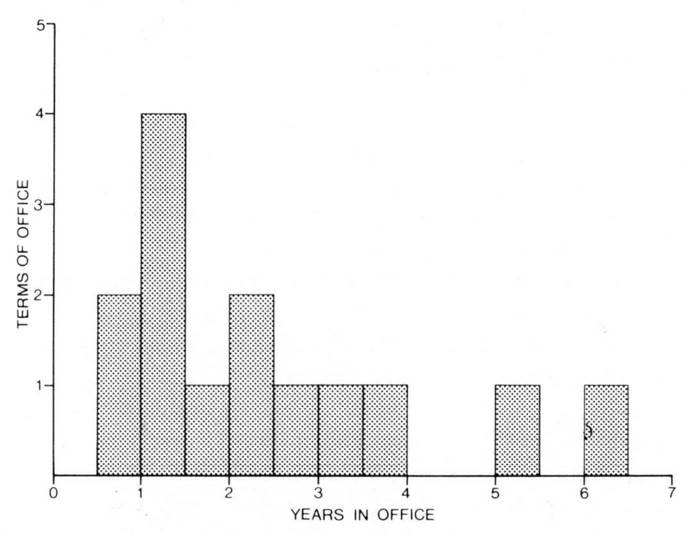

FIG. 14.1. Length of ministerial office: shown is the distribution of terms of office by transport "ministers" according to the length of time each term lasted. Six terms between 1945–79 were for a period of less than eighteen months.

But such an implication and one "Peeler" do not constitute a case and, as Lord Crowther-Hunt reminded his readers, "any number of assertions can be made about the growth of civil service power – but they remain assertions no matter who is making them".

In contrast to the mandarins, the road and environment lobbies were distinguishable by pursuit of overt objectives. At different times the roads lobby or the environment lobby appeared to hold the centre of the stage, and they undoubtedly worked hard to influence road and traffic policies. The arguments presented by some groups, for example, were well researched and well reasoned. But whether, as a consequence, road and traffic policies were simply the outcome of "a symbiotic relationship between groups and government . . ."[15] is a moot point. It is interesting to note that Plowden, for example, in his fundamental review of motor-car policy tended to play down the influence of interest groups. Thus,

> "to explain official policy as a whole as the product of successful lobbying would be to misunderstand the character of British government and the ways in which decisions on policy are . . . made."

He added:

> "even in the single area in which all motor interests see eye to eye – roads themselves – it cannot be said that governments have been persuaded to invest more heavily than the [public] demand for road space warrants."[16]

This view aside, there are certain features normally associated with the pressure group thesis that appear to be absent in the context of changing road and traffic policies. The pressure group thesis tends to emphasize the viewpoint that a change in policy emerges from a bargaining process between rival groups, with the government holding the ring and acting as a form of referee. Consequently, in this context the typical adjustment to policy is seen as a small-scale, often rather muddled, adjustment. But it has been pointed out already that principal features of road and traffic policies since the Second World War were periods of long-term stability punctuated by a few occasions when change in policy was rather distinct and positive.

Thus, there are reasons for sharing Plowden's scepticism of the view that democracy has passed beyond the parliamentary stage and that the formulation of policy is now exercised solely by colluding group interests. In fact, probably the most telling reason for rejecting this view in the present context is the very evidence that the pressure group theorists would point to – the Peeler note. The note contained the following phrase:

> ". . . it is assumed that we wish . . . to move, *as soon as parliamentary and public opinion will let us*, to a maximum gross weight of thirty-eight or forty tons"[17] (present writer's emphasis).

If the Peeler memorandum indicated a "departmental view" in consort with the road-haulage lobby, it suggested also that parliamentary democracy was alive and very capable of kicking back. Therefore, in explaining the nature of change in road and traffic policies since the Second World War it appears that we need to take a broader viewpoint.

Force of Circumstance

Important here have been secular trends in the economy or what Richard Rose has referred to as the "force of circumstance" working through and influencing opinion.[18] For most of the fifties and during the early years of the sixties, the rate of growth in disposable incomes was, by later standards, relatively high. There was also an expectation that this performance would be repeated, if not subsequently improved upon. Such high expectations fed through into road and traffic policies. They were manifest, for example, in the assumed high rates of growth in car ownership (and therefore traffic) used in the transport studies during the sixties decade. In turn, the combination of this expected pressure on road space and the expectation that the country would be able to afford higher levels of public investment, led naturally to an emphasis on road building.

Such tendencies and inclinations were further reinforced by other circumstantial factors. Important here was the visionary stimulus provided at a most opportune time by the Buchanan Report. In addition, in the early sixties the country was in the throes of a mania for comprehensive redevelopment of cities fuelled by a boom in commercial property. This provided yet another opportunity for large-scale urban road building. Thus in the sixties a number of powerful factors combined in such a way that an urban policy, sharply focused on the prospect of new, large-scale highways, was the natural outcome.

In the seventies this process of an accumulation of circumstantial causes was to be seen working in an opposite direction. The rate of growth in the economy was lower, future supplies of petroleum were regarded as uncertain, attitudes to public investment had changed and the ethos of the time now stressed conservation and rehabilitation rather than redevelopment. Again policies for managing the demand for existing road space flowed naturally from such an environment.

The same ingredients of secular economic and social change combining to either weaken or strengthen, confirm or overturn existing policies, were to be seen also in the context of trunk roads and heavy lorries. Thus, whilst the idea of urban motorways had declined with the fortunes of comprehensive redevelopment, the trunk-road programme was sustained into the seventies by the continued existence of regional problems and the maintenance of the strongly held belief that motorways would have a catalytic effect upon such ailing regional economies.

Similarly, a powerful contributory factor leading to the abandonment of the heavy-lorry network in 1977 was the economic recession. Economic pressures on the road-haulage industry in the latter half of the seventies were such that the arguments that a routeing system would impose unreasonable penalties on the industry received a more sympathetic hearing. At the same time cuts in public expenditure, and consequently in the roads programme, were making it more difficult to fulfil the aim of building a plausible heavy-lorry network by the early 1980s.

A Synthesis

However, such secular economic and social trends, which on the whole were of a fairly smooth nature, do not by themselves explain one important feature of post-war road and traffic policies, namely the tendency towards distinct, quite sudden but infrequent change. However, by combining the forces of circumstance with the role of pressure groups, we are able to explain more easily such marked changes in policy.

Just as pressure groups promote their preferred policies so too do they resist change which either is not in their interest or with which they do not associate. Mandarins in particular act in this manner. The Whitehall style of government is to avoid conflict and seek consensus. The inclination of the civil servant is to advise in favour of change only when an assured position has been reached with most, if not all, interested parties (including here the perceived position of the electorate viewed as a whole). Thus, "political caution and administrative inertia"[19] in the Whitehall system, operate to deflect challenge to existing policies.

The deflection by Whitehall of pressures from economists lobbying for road pricing is a case in point. The early reaction of the civil servants to the economists' case was one of caution. It was a reaction assisted initially, it is claimed, by the inability of the protagonists to present their case in a convincing way, and, later, by a slower than expected growth in traffic congestion.[20] It was a challenge to the status quo that was met successfully.

In these circumstances, alteration to policies is unlikely to be brought about by gradual modification. On the contrary, the tendency is for a firm change of policy to take place only when a sufficient head of steam has been produced by lobbyists capitalizing upon favourable pressures from economic, social and intellectual forces. With sufficient push the old policies will give way, thus producing a distinctive break with the past.* It is possible, of course, that such a powerful consensus against existing policies fails to focus upon a generally agreed new course of action. If this is the case there will be a tendency for policy to regress towards Plowden's depiction of a "non-policy". For example, the lorry-routeing saga illustrates such a tendency. But, on the basis of what has normally happened during the last thirty-five years, the usual transmogrification will be from one positive policy to another; from building urban motorways, for example, to managing the demands for existing road space.

In the late seventies one could detect "forces of circumstance" combining to increase pressures on existing urban road and traffic policies. The external factors in this instance were manifest in an inner-city problem and the belief that better road communications to and from such areas could stem or reverse the tide of declining jobs. Consequently, there were growing calls for increased spending on inner-city roads and for a relaxation of parking restraint to help inner-city firms. In addition, there were more general pressures for increased road expenditure in order to provide the national economy with a much needed boost. Whether in the future these forces of change will coalesce and strengthen to such a degree that a new, distinct urban roads policy emerges remains to be seen.

But if road spending in general does increase significantly again in the near future it is unlikely to approach the high levels witnessed in the late sixties and early seventies. Nor is it likely that higher road expenditure will be manifest

*This suggests that the policy process may be susceptible to rigorous analysis using techniques of *catastrophe theory*. Catastrophe theory is a new branch of mathematics which depicts the mechanism whereby gradual variation in a set of factors results in an abrupt, or sudden, change in a state of affairs. One facet of this theory deals with the so-called hysteresis effect; if the pressures which have led to a (sudden) change are reversed, then a similar reversal of the state of affairs will not occur at the same level of pressure. The significance of the hysteresis effect is that it suggests in an environment such as this, where administrative and political inertia is important, that policies will change in accordance with a particular pattern.

in an intensive programme of motorway construction. Proposals for motorway networks were Britain's peculiar answer to a problem defined in a particular way at an appropriate stage in its history. Those days have gone forever – and with them the "motorway age".

References

Preface

1. T. P. HUGHES, "Roads Policy at National Regional and Local Levels and the Role of Motorways", in *20 Years of British Motorways,* Institution of Civil Engineers, London, 1980.
2. JOHN TYME, *Motorways Versus Democracy,* Macmillan, London, 1978.

Chapter 1: THE FIRST MOTORWAYS

1. *The Times,* 6 December 1958 ("Prime Minister breaks rules: a closer look at first motorway").
2. HOUSE OF COMMONS, *Debates,* **422,** Col. 590–5 (6 May 1946).
3. JOHN BOYD-CARPENTER, *Way of Life,* Sidgwick & Jackson, London, 1980, p. 107.
4. HOUSE OF COMMONS, *Debates,* **422,** Col. 590–5 (6 May 1946).
5. JOHN BOYD-CARPENTER, *Way of Life,* Sidgwick & Jackson, London, 1980, p. 111.
6. HOUSE OF COMMONS, *Debates,* **574,** Col. 48 (22 July 1957).
7. HOUSE OF COMMONS, *Trunk Roads,* First Report from the Select Committee on Estimates, Session 1958/59, HMSO, London.
8. T. M. COBURN, M. E. BEESLEY and D. J. REYNOLDS, "The London–Birmingham Motorway: Traffic and Economics", *Road Research Laboratory Technical Paper* 46, HMSO, London, 1960.

Chapter 2: TOWN ROADS: A CONSERVATIVE ERA

1. PETER HALL, *Urban and Regional Planning,* Penguin Books, Harmondsworth, 1974, p. 114.
2. Sir FREDERICK COOK (Chairman), *Design and Layout of Roads in Built-up Areas,* Report of the Departmental Committee set up by the Minister of War Transport, HMSO, London, 1946.
3. Quoted in D. S. THOMAS, "Birmingham Inner Ring Road – Costs, Savings and Benefits", *People and Cities Conference,* British Roads Federation, London, 1963.
4. Quoted in C. M. BUCHANAN, "London Road Plans 1900–1970", *Greater London Research Intelligence Unit Research Report* 11, GLC, London, 1970, p. 35.
5. Adapted from DEPARTMENT OF SCIENTIFIC AND INDUSTRIAL RESEARCH, *Road Research 1953,* HMSO, London, 1953, Table II.

6. ALAN DAY, *Roads,* Mayflower Books, London, 1963, p. 14.
7. WILLIAM PLOWDEN, *The Motor Car and Politics in Britain,* Penguin Books, Harmondsworth, 1973, p. 362.
8. W. H. GLANVILLE and J. F. A. BAKER, "Urban Motorways in Great Britain?", *Urban Motorways, Report of the London Conference,* British Roads Federation, London, 1956, p. 23.

Chapter 3: SQUEEZING MORE OUT OF EXISTING STREETS

1. HAROLD MACMILLAN, *Pointing the Way 1959–1961,* Macmillan, London, 1972, p. 19.
2. *Report of the Royal Commission on Local Government in Greater London* (Cmnd. 1164), HMSO, London, 1960, para. 406.
3. HOUSE OF COMMONS, *Debates,* **614,** Col. 560 (26 November 1959).
4. *The Guardian,* 1 December 1959.
5. *The Guardian,* 8 December 1959.
6. ALAN DAY, *Roads,* Mayflower, London, 1963, pp. 22 and 60.
7. "Transport is Everyone's Problem", *Socialist Commentary, Special Supplement* (April 1963) p. xxvii.
8. *The Guardian,* 21 November 1964 ("More one-way roads forecast in towns").
9. HOUSE OF COMMONS, *Debates,* **689,** Col. 32, 10 February 1964.
10. PETER HALL, "Implementing Buchanan – Short Term Policies", *Traffic Engineering and Control,* **5** (1964) 702–5.
11. HOUSE OF COMMONS, *Debates,* **689,** Col. 32–3 (10 February 1964).

Chapter 4: THE BUCHANAN REPORT: TRAFFIC IN TOWNS

1. HOUSE OF COMMONS, *Debates,* **615,** Col. 771 (10 December 1959).
2. C. D. BUCHANAN, *Mixed Blessing: The Motor in Britain,* Leonard Hill, London, 1958.
3. Sir R. CLARKE, *Public Expenditure, Management and Control,* Macmillan, London, 1978, para. 48.
4. C. D. BUCHANAN *et al., Traffic in Towns: A Study of the Long Term Problems of Traffic in Urban Areas,* Reports of the Steering and Working Groups appointed by the Minister of Transport, HMSO, London 1963.
5. C. D. BUCHANAN, "Standards and Values in Motor Age Towns", *Journal of the Town Planning Institute,* **47** (1961) 320–9.
6. HOUSE OF COMMONS, *Debates,* **615,** Col. 763 (10 December 1959).
7. C. D. FOSTER, "Estimates of the Cost of Buchanan", *Statist,* **183** (1964) 546–51.
8. W. G. SMIGIELSKI *et al., Leicester Traffic Plan,* City of Leicester, 1964.
9. *The Observer,* 8 November 1964 (D. H. CROMPTON, "How to keep the cars away").
10. M. E. BEESLEY and J. F. KAIN, "Urban Form, Car Ownership and Public Policy: An Appraisal of 'Traffic in Towns'", *Urban Studies,* **1** (1964) 174–203. See also J. F. KAIN and M. E. BEESLEY, "Forecasting Car Ownership and Use", *Urban Studies,* **2** (1965) 163–85.
11. HOUSE OF COMMONS, *Debates,* **689,** Col. 37 (10 February 1964).
12. HOUSE OF COMMONS, *Debates,* **689,** Col. 364 (12 February 1964).
13. ROADS CAMPAIGN COUNCIL, *Roads Ready Reckoner,* Roads Campaign Council, London, 1964.
14. MINISTRY OF TRANSPORT, *Roads Circular No.* 8, 1965.

15. STANDING CONFERENCE ON LONDON REGIONAL PLANNING, *Road Construction and Improvement Prospects in the Conference Area,* Report by the Technical Panel, 1966.
16. D. J. REYNOLDS, *Economics, Town Planning and Traffic,* Institute of Economic Affairs, London, 1966.

Chapter 5: THE ROAD-PRICING DEBATE

1. In *Urban Motorways, Report of the London Conference,* British Roads Federation, London, 1956, p. 53.
2. In *Urban Motorways, Report of the London Conference,* British Roads Federation, London, 1956, p. 56.
3. C. D. BUCHANAN, *Mixed Blessing: The Motor in Britain,* Leonard Hill, London, 1958, p. 197.
4. G. ROTH, *Paying for Roads: The Economics of Traffic Congestion,* Penguin Books, Harmondsworth, 1967, p. 34–5.
5. See for example:
   G. ROTH, "A Pricing Policy for Road Space in Town Centres", *Journal of the Town Planning Institute,* **47** (1961) 287–9.
   A. A. WALTERS, "The Theory and Measurement of Private and Social Cost of Highway Congestion", *Econometrica,* **29** (1961) 679–99.
   M. E. BEESLEY and G. J. ROTH, "Restraint of Traffic in Congested Areas", *Town Planning Review,* **33** (1962) 184–96.
   G. ROTH and J. M. THOMSON, "Road Pricing: A Cure for Congestion?", *Aspect,* **1** (1963) 3.
6. "Transport is Everyone's Problem", *Socialist Commentary, Special Supplement* (April 1963).
7. G. ROTH, "An End to Traffic Jams", *Crossbow* (July–September 1962).
8. R. SMEED (Chairman), *Road Pricing: The Economic and Technical Possibilities,* HMSO, London, 1964.
9. G. F. RAINNIE, "Review of 'Traffic in Towns' ", *Journal of the Royal Statistical Society,* **131** (Series A) (1968) 75–83.
10. HOUSE OF COMMONS, *Debates,* **689,** Col. 36 (10 February 1964).
11. HOUSE OF COMMONS, *Debates,* **689,** Col. 364 (12 February 1964).
12. Reported in *The Times,* 11 June 1964 ("Group to study design in vehicles").
13. Reported in *The Guardian,* 1 April 1965 ("Minister foresees radical measures in traffic control").
14. Reported in *The Times,* 15 April 1965 ("Higher tax for city cars?").
15. Reported in *The Times,* 7 October 1966 ("Mrs. Castle learns from U.S. mistakes").
16. Reported in *The Times,* 11 December 1965 ("Mr. Du Cann says government is anti-motorist").
17. Reported in *The Times,* 11 December 1965 ("Mr. Du Cann says government is anti-motorist").
18. Reported in *The Times,* 19 October 1966 (p. 13).
19. Reported in *The Times,* 4 November 1966 (p. 12).
20. MINISTRY OF TRANSPORT, *Better Use of Town Roads,* HMSO, London, 1967.
21. Reported in *The Times,* 1 April 1965 ("Minister foresees radical measures on traffic control").
22. C. D. BUCHANAN, "Meeting the Challenge of the Car", *The Times,* Special Report on Roads, 19 September 1966.
23. Reported in *The Times,* 4 November 1966 (p. 12).
24. Reported in *The Guardian,* 7 July 1966 ("Car meters plan 'too simple' ").

Chapter 6: BRIDGING THE GAP

1. MINISTRY OF TRANSPORT, *Transport Policy* (Cmnd. 3057), HMSO, London, 1966, para. 50.
2. P. G. GRAY, *Private Motoring in England and Wales,* Government Social Survey, HMSO, London, 1969.
3. MINISTRY OF TRANSPORT AND MINISTRY OF HOUSING AND LOCAL GOVERNMENT, "Parking in Town Centres", *Planning Bulletin* 7, HMSO, London, 1965.
4. MINISTRY OF TRANSPORT, "Traffic and Transport Plans", *Roads Circular* 1/68, HMSO, London, 1968.
5. A. D. MAY, "Parking Control for Restraint in Greater London", *GLC Intelligence Unit Quarterly Bulletin* 19, June 1972.
6. D. GREENWOOD "Urban Congestion Study 1976: Interim Report", *Traffic Advisory Unit Report,* Department of Transport, 1978.
7. A. D. MAY, "Comment: Road Pricing – Should and Might It Happen?" *Transportation,* 8 (1979) 119–23.
8. Lord KINGS NORTON (Chairman), *Cars for Cities: A Study of Trends in the Design of Vehicles with Particular Reference to their Use in Towns,* Reports of the Steering and Working Group, HMSO, London, 1967.
9. Reported in *The Times,* 11 June 1964 ("Group to study design in vehicles: Minister hints at smaller cars").
10. T. H. BENNETT, "The Physical Characteristics of the British Motor Vehicle Fleet", *The Highway Engineer* (October 1976) 24–9.
11. MINISTRY OF TRANSPORT, *Public Transport and Traffic* (Cmnd. 3481), HMSO, London, 1967, para. 115.

Chapter 7: "SCIENTIFIC" ROAD PLANS AND URBAN RENEWAL

1. MINISTRY OF TRANSPORT, *Transport Policy* (Cmnd. 3057), HMSO, London, 1966, para. 50.
2. *The Financial Times,* 26 November 1964 ("Are road plans going in the right direction?").
3. HOUSE OF COMMONS, *Debates,* **689,** Col. 45 (10 February 1964).
4. HOUSE OF COMMONS, *Debates,* **689,** Col. 39 (10 February 1964).
5. HOUSE OF COMMONS, *Urban Transport Planning,* Second Report from the Expenditure Committee, Session 1972/73, HMSO, London, para. 149.
6. W. KEEGAN and R. PENNANT-REA, *Who Runs the Economy?: Control and Influence in British Economic Policy,* Temple Smith, London, 1979.
7. See for example COLIN BUCHANAN and Partners, *The Conurbations,* British Roads Federation, London, 1969.
8. HOUSE OF COMMONS, *Debates,* **689,** Col. 148 (10 February 1964).
9. C. D. BUCHANAN, "Comprehensive Redevelopment – The Opportunity for Traffic", in T. E. H. WILLIAMS (editor), *Urban Survival and Traffic,* Spon, London, 1962.
10. *The Guardian,* 22 January 1964 (BRIAN REDHEAD: "Councils ignore Buchanan lessons").
11. MINISTRY OF TRANSPORT AND MINISTRY OF HOUSING AND LOCAL GOVERNMENT, Joint Circular, "Traffic In Towns", *Circular* 1/64, 1964.
12. *The Guardian,* 30 December 1964 ("Roads plans 'did not go far enough'").
13. MINISTRY OF TRANSPORT, *Roads in England,* HMSO, London, 1966, p. 14 and Appendix to Chapter 5.
14. Lady SHARP, *Transport Planning: The Men for the Job, Report to the Minister of Transport,* HMSO, London, 1970.

15. B. V. MARTIN, "London Transportation Study: A Review", in *Proceedings, Transportation Engineering Conference,* Institute of Civil Engineers, London, 1968, pp. 1–11.
16. ECONOMIST INTELLIGENCE UNIT, "Urban Transport Planning in the United Kingdom", *Motor Business,* **40** (1964) 17–32.
17. M. E. BEESLEY *et al.*, "Urban Transport Models and Motorway Investment", *Economica,* **30** (New Series) (1963) 243–61.
18. S. PLOWDEN, "Transportation Studies Examined", *Journal of Transport Economics and Policy,* **1** (1967) 5–27.
19. A. M. VOORHEES, *Traffic in the Conurbations,* British Roads Federation, London, 1971, p. 9.
20. See D. N. M. STARKIE, *Transportation Planning and Public Policy,* Pergamon, Oxford, 1973, p. 380.
21. D. A. QUARMBY, "Choice of travel mode for the journey to work: Some findings", *Journal of Transport Economics and Policy,* **1** (1962) 273–314.
22. HOUSE OF COMMONS, *Urban Transport Planning,* Second Report from the Expenditure Committee, Session 1972/3 HMSO, London, para. 162.
23. P. W. J. BATEY and J. V. NICKSON, "Transport Modelling as an Aid in Structural Planning: A New Lease of Life for the SELNEC Model?", *Working Paper* 3, Department of Civic Design, University of Liverpool, 1976.
24. See J. S. DODGSON, "The Development of Transport on Merseyside since 1945", in B. L. ANDERSON and P. J. M. STONEY (editors), *Commerce, Industry and Transport: Studies in Economic Changes on Merseyside,* Liverpool University Press, 1981.
25. C. M. BUCHANAN, "London Road Plans 1900–1970", *Greater London Research Intelligence Unit Research Report* 11, GLC, London, 1970.
26. *Glasgow's Transport Plan – Dream or Reality?,* Corporation of Glasgow, 1969.
27. ANON., "City on the Move", *The Economist,* **213** (1964) 605.
28. S. E. EVANS and I. H. MACKINDER, "Predictive Accuracy of British Transportation Studies", *Conference Proceedings,* PTRC, London, 1980.
29. JOHN BLAKE, "The Growth of Traffic in Greater London: Buchanan and London Traffic Surveys Too Alarmist?", *Surveyor – Local Government Technology,* 19 October 1968 pp. 34–5.
30. A. W. EVANS, "Myths About Employment in London", *Journal of Transport Economics, Policy,* **1** (1967) 214–25.
31. HOUSE OF COMMONS, *Urban Transport Planning,* Second Report from the Expenditure Committee, Session 1972/73, HMSO, London, Vol. I, para. 153.
32. J. V. CABLE, "Glasgow's Motorways: A Technocratic Blight", *New Society,* **30** (1974) 605–7.

Chapter 8: URBAN ROADS: THE ENVIRONMENTAL BACKLASH

1. HOUSE OF COMMONS, *Urban Transport Planning,* Second Report from the Expenditure Committee, Session 1972/3, HMSO, London, Vol. II, p. 25.
2. BRITISH ROADS FEDERATION, *Urban Motorways, Report of the London Conference,* British Roads Federation, London 1956.
3. F. BAKER, *Riverside Highway – the Evening News Plan,* Associated Newspapers, London, 1956.
4. Quoted in C. M. BUCHANAN, "London Road Plans 1900–1970", *Greater London Research Intelligence Unit Research Report* 11, GLC, London, 1970.
5. A. DAY, *The Observer,* 29 January 1961.
6. Quoted in D. A. HART, *Strategic Planning in London: The Rise and Fall of the Primary Road Network,* Pergamon, Oxford, 1976. p. 112.
7. "Transport is Everyone's Problem", *Socialist Commentary, Special Supplement* (April 1963).

8. J. BARR, "Bath and Buchanan", *New Society,* **7** (1966) 192, 5–7.
9. S. PLOWDEN, *Towns Against Traffic,* Andre Deutsch, London, 1972.
10. MINISTRY OF TRANSPORT, *Roads in England and Wales 1965,* HMSO, London, 1965, para 2.20.
11. PETER HALL, "Where London's Roads Will Go", *New Society,* **5** (1965) 130, 18–19.
12. *Evening Standard,* 25 March 1965 ("First map of the new urban motorway").
13. D. A. HART, *Strategic Planning in London: The Rise and Fall of the Primary Road Network,* Pergamon, Oxford, 1976.
14. P. G. HALL, *Great Planning Disasters,* Weidenfeld & Nicolson, London, 1980.
15. *Evening Standard,* 31 March 1965.
16. *Evening Standard,* 3 June 1967.
17. *Evening Standard,* 1 November 1967.
18. *The Sunday Times,* 7 May 1967.
19. *Evening News,* 14 March 1969 ("London's motorways: is there another answer?").
20. *The Times,* 8 January 1969 (M. BAILY, "Problems of cutting across the face of a city").
21. TONY ALDOUS, *Battle for the Environment,* Fontana, London, 1972, p. 44.
22. JOHN BARR, "Divided City", *New Society,* **13** (1969) 701–4.
23. J. GRANT, *The Politics of Urban Transport Planning,* Earth Resources, London, 1974.
24. MINISTRY OF TRANSPORT, *Roads for the Future: a New Inter-Urban Plan,* HMSO, London, 1969.
25. TREASURY, *Public Expenditure to 1975–76* (Cmnd. 4829), HMSO, London, 1971.
26. M. E. BEESLEY and J. F. KAIN, "Urban Form, Car Ownership and Public Policy: an Appraisal of the Traffic in Towns", *Urban Studies,* **1** (1974) 174–203.
27. DEPARTMENT OF THE ENVIRONMENT, *New Roads in Towns,* HMSO, London, 1972.
28. DEPARTMENT OF THE ENVIRONMENT, *Development and Compensation – Putting People First* (Cmnd. 5124), HMSO, London, 1972.
29. DEPARTMENT OF THE ENVIRONMENT, "Noise Insulation Regulations 1975", *Circular* 114/75, HMSO, London, 1975.
30. A. ALEXANDRE *et al.*, "The Economics of Traffic Noise Abatement", *Traffic Quarterly* (April 1976) 269–83.
31. HOUSE OF COMMONS, *Urban Transport Planning,* Second Report from the Expenditure Committee, Session 1972/73, HMSO, London, Vol. II, pp. 1–2.

Chapter 9: TOWN TRAFFIC AND ROADS: WINDS OF CHANGE

1. J. M. THOMSON *et al., Motorways in London,* Gerald Duckworth, London, 1969.
2. DEPARTMENT OF THE ENVIRONMENT, *Greater London Development Plan: Report of the Panel of Inquiry,* HMSO, London, 1973.
3. P. G. HALL, *Great Planning Disasters,* Weidenfeld & Nicolson, London 1980 (see Chapter 3, "London's Motorways").
4. *The Guardian,* 27 January 1968 ("Coventry planning to give buses priority over cars during rush hours").
5. HOUSE OF COMMONS, *Urban Transport Planning,* Second Report from the Expenditure Committee, Session 1972/73, HMSO, London.
6. *Traffic in Towns,* Report of the (Crowther) Steering Group appointed by the Minister of Transport, HMSO, London 1963.
7. *The Financial Times,* 26 November 1964 ("Are road plans going in the right direction?").
8. HOUSE OF COMMONS, *Debates,* **853**, Col. 323–76 (20 March 1973).

9. HOUSE OF COMMONS, *Debates,* **859,** Col. 1047–1119 (9 July 1973).
10. DEPARTMENT OF THE ENVIRONMENT, *Urban Transport Planning: Government Observations on the Second Report of the Expenditure Committee* (Cmnd. 5366), HMSO, London, 1973.
11. DEPARTMENT OF THE ENVIRONMENT, "Local Transport Grants", *Circular* 104/73, HMSO, London, 1973.
12. HOUSE OF COMMONS, *Debates,* **865,** Col. 399 (28 November 1973).
13. See HOUSE OF COMMONS, *Public Expenditure and Transport,* First Report from the Expenditure Committee, Session 1974, HMSO, London, p. 8, para. 2.

Chapter 10: URBAN TRAFFIC AND ROADS: POST-1973

1. DEPARTMENT OF THE ENVIRONMENT, "Transport Supplementary Grant Submissions for 1976/7", *Circular* 43/75, HMSO, London, April 1975.
2. D. J. LYONS, "Road Pricing", in P. M. TOWNROE (editor), *Social and Political Consequences of the Motor Car,* David & Charles, Newton Abbot, 1974.
3. T. L. BEAGLEY (Chairman), *Report to the Minister, Working Group on Bus Demonstration Projects,* Ministry of Transport, 1970.
4. HOUSE OF COMMONS, *Public Expenditure on Transport,* First Report from Expenditure Committee, Session 1974, HMSO, London.
5. DEPARTMENT OF TRANSPORT, *Transport Policy* (Cmnd. 6836), HMSO, London, 1977, para. 123.
6. DEPARTMENT OF THE ENVIRONMENT, *Transport Policy: A Consultation Document,* HMSO, London, 1976, Vol. 2, para. 3.17.
7. DEPARTMENT OF THE ENVIRONMENT, "A Study of Some Methods of Traffic Restraint", *Research Report* 15, HMSO, London, 1976.
8. R. A. VINCENT and R. E. LAYFIELD, "Nottingham Zones and Collar Study – Overall Assessment", *Report* LR 805, Transport and Road Research Laboratory, Crowthorne, 1977.
9. *Daily Telegraph,* 18 November 1977 ("Boroughs to oppose curb on parking").
10. DEPARTMENT OF THE ENVIRONMENT, "Transport Supplementary Grant Submissions for 1977/78", *Circular* 125/75, HMSO, London, December 1975, para. 24.
11. DEPARTMENT OF THE ENVIRONMENT, "York Inner Ring Road – Decision announced", *Press Notice* 91, HMSO, London, 1965.
12. *The Times,* 18 August 1978 ("Oxford road scheme rejected by Minister").
13. See also J. M. BAILEY, "The Evolution of Transport Policy in Oxford", in T. ROWLEY (editor), *The Oxford Region,* Oxford University Department of External Studies, 1980.
14. HOUSE OF COMMONS, *The Roads Programme,* First Report from the Transport Committee, Session 1980/81, HMSO, London, p. 43.
15. T. P. HUGHES, "Roads Policy at National, Regional and Local Levels and the Role of Motorways", in *20 Years of British Motorways,* Institution of Civil Engineers, London, 1980.
16. P. G. HALL, "Monumental Folly", *New Society,* **12** (1968) 602–3.
17. A. STONES, "Stop Slum Clearance – Now", *Official Architecture and Planning,* **35** (1972) 107–110.
18. See R. BOTHAM and J. HERSON, "Motorway Madness in Liverpool? – From Rationality to Rationalisation", *The Planner,* **66** (1980) 118–21.

Chapter 11: THE JUGGERNAUT LORRY ISSUE

1. A. J. HARRISON, "Road Transport Costs", *Oxford University, Institute of Economics and Statistics Bulletin,* 27 (1965) 103–117.

2. Lord GEDDES (Chairman), *Carriers' Licensing,* Report of the Committee, Ministry of Transport, HMSO, London, 1965.
3. R. KIMBER, J. J. RICHARDSON and S. K. BROOKES, "The Juggernauts: Public Opposition to Heavy Lorries", in R. KIMBER and J. J. RICHARDSON (editors), *Campaigning for the Environment,* Routledge & Kegan Paul, London, 1974.
4. HOUSE OF COMMONS, *Debates,* **788** (20 October 1969).
5. CIVIC TRUST, *Heavy Lorries,* London, 1970.
6. HOUSE OF COMMONS, *Debates,* **847**, Col. 511–60 (29 November 1972).
7. TRANSPORT 2000, *Weighing the Evidence,* London 1979.
8. P. J. CORCORAN, M. H. GLOVER and B. A. SHANE, "Higher Gross Weight Goods Vehicles – Operating Costs and Road Damage Factors", *Supplementary Report* 590, Transport and Road Research Laboratory, Crowthorne, 1980, p. 7.
9. Sir ARTHUR ARMITAGE, *Report of the Inquiry into Lorries, People and the Environment,* HMSO, London, 1980.
10. DEPARTMENT OF TRANSPORT, *Lorries, People and the Environment: a Background Paper,* September 1979.
11. FREIGHT TRANSPORT ASSOCIATION, *Freight Facts: The Case for the Heavier Lorry,* FTA, 1979.
12. Reported in *The Financial Times,* 5 December 1980 ("Hauliers expect recovery will not begin until next year").
13. C. D. FOSTER (Chairman), *Road Haulage Operators' Licensing,* Report of the Independent Committee of Inquiry, HMSO, London, 1979, Table B.20.

Chapter 12: LORRY ROUTES AND ROAD PLANS

1. *Evening Standard,* 30 July 1964 ("I won't make lorry men use M1 says Marples").
2. MINISTRY OF TRANSPORT, *Roads for the Future: a New Inter-urban Plan,* HMSO, London, 1969.
3. MINISTRY OF TRANSPORT, *Roads for the Future* (Cmnd. 4369), HMSO, London, 1970.
4. HOUSE OF COMMONS, *Debates,* **819**, Written Answers, Col. 288–9 (23 June 1971).
5. CIVIC TRUST, *Heavy Lorries,* preliminary memorandum, 1970.
6. HOUSE OF COMMONS, *Debates,* **836**, Oral Answers, Col. 1338 (24 November 1971).
7. DEPARTMENT OF THE ENVIRONMENT, *Press Notice* 230m, 7 October 1971 ("Making the heavy lorry 'more civilized'").
8. DEPARTMENT OF THE ENVIRONMENT, *Press Notice* 256, 26 April 1971 ("heavier lorries possible but restricted to suitable roads says Transport Minister").
9. DEPARTMENT OF THE ENVIRONMENT, *Press Notice* 556m, 9 May 1972 ("'Come to terms with the environment': Minister of Transport Industries' call to road hauliers").
10. DEPARTMENT OF THE ENVIRONMENT, *Press Notice* 615m, 24 May 1972 ("Special lorry routes may be inevitable, says Mr. Peyton").
11. DEPARTMENT OF THE ENVIRONMENT, *Press Notice* 1100m, 4 October 1972 ("New techniques for routeing lorries – Mr. John Peyton").
12. *The Daily Telegraph,* 24 September 1964 ("Scheme to ban heavy lorries on minor roads").
13. HOUSE OF COMMONS, *Debates,* **849**, Col. 1786–1850 (2 February 1973).
14. DEPARTMENT OF THE ENVIRONMENT, *Routes for Heavy Lorries: a Consultation Paper,* July 1974.
15. DEPARTMENT OF THE ENVIRONMENT, *Routes for Heavy Lorries: a Consultation Paper,* July 1974, p. 4.
16. CIVIC TRUST, *Heavy Lorries: Nine Years On,* London, 1979, p. 16.
17. P. BLAKE, *Heavy Lorry Control, Report* T92, Local Government Operational Research Unit, Reading, 1980, p. 37.

18. DEPARTMENT OF TRANSPORT, *Transport Policy* (Cmnd. 6836), HMSO, London, 1977.
19. FREIGHT TRANSPORT ASSOCIATION, *Five Years of Lorry Management Since Dykes,* Tunbridge Wells, Kent, 1978.
20. Sir A. ARMITAGE, *Report of the Inquiry into Lorries, People and the Environment,* HMSO, London, 1980.

Chapter 13: "THE LAST MOTORWAY?"

1. JOHN TYME, *Motorways Versus Democracy,* Macmillan, London, 1978.
2. The saga is well covered in ROY GREGORY, "The Minister's Line: or, the M4 comes to Berkshire", *Public Administration,* **45** (1967) 113–128 and 269–86.
3. Quoted in P. H. LEVIN, "Highway Inquiries: a Study in Government Responsiveness", *Public Administration,* **57** (1979) 21–49.
4. HOUSE OF COMMONS, *Public Expenditure on Transport,* First Report from the Expenditure Committee, Session 1974, HMSO, London, para. 59.
5. HOUSE OF COMMONS, *Observations by the Department of the Environment on the First Report from the Expenditure Committee in Session 1974,* Fourth Special Report from the Expenditure Committee Session 1974/75, HMSO, London, para. 9.
6. Summarized in P. H. LEVIN, "Highway Inquiries: a Study in Government Responsiveness", *Public Administration,* **57** (1979) 21–49.
7. J. C. TANNER, "Forecasts of Future Numbers of Vehicles in Great Britain", *Roads and Road Construction,* **40** (1962) 263–74.
8. Sir GEORGE LEITCH (Chairman), *Report of the Advisory Committee on Trunk Road Assessment,* Department of Transport, HMSO, London, 1978.
9. DEPARTMENT OF TRANSPORT *et al., Report on the Review of Highway Inquiry Procedures* (Cmnd. 7133), HMSO, London, April 1978.
10. DEPARTMENT OF TRANSPORT, *Policy for Roads: England 1978* (Cmnd. 7132), HMSO, London.
11. DEPARTMENT OF TRANSPORT, *Policy for Roads: England 1980* (Cmnd. 7908), HMSO, London.
12. KENNETH CLARKE, "Opening Address", in *20 Years of British Motorways,* Institution of Civil Engineers, London, 1980.
13. T. P. HUGHES, "Roads Policy at National, Regional and Local Levels and the Role of Motorways", in *20 Years of British Motorways,* Institution of Civil Engineers, London, 1980.
14. HOUSE OF COMMONS, *The Roads Programme,* First Report from the Transport Committee, Session 1980/81, HMSO, London.
15. HOUSE OF COMMONS, *Motorways and Trunk Roads,* Sixth Report of the Estimates Committee, Session 1968/69, HMSO, London.
16. BOARD OF TRADE, *The North-East: A Programme for Regional Development and Growth* (Cmnd. 2206), HMSO, London, 1963.

Chapter 14: PERSPECTIVES ON THE POLICY PROCESS

1. WILLIAM PLOWDEN, *The Motor Car and Politics in Britain,* Penguin Books, Harmondsworth, 1973.
2. K. CLARKE, "Opening Address", in *20 Years of British Motorways,* Institution of Civil Engineers, London, 1980.
3. HOUSE OF COMMONS, *Government Observations on the First Report of the Committee, Session 1980/81,* Third Special Report from the Transport Committee Session 1980/81, HMSO, London.

4. See for example J. E. S. HAYWARD, "National Aptitudes for Planning in Britain, France and Italy", *Government and Opposition,* 9 (1974) 397–410.
5. HOUSE OF COMMONS, *Debates,* 847, Col. 511–60 (29 November 1972).
6. D. BRAYBROOKE, *Traffic Congestion Goes Through the Issue-Machine: a Case Study in Issue Processing, Illustrating a New Approach,* Routledge & Kegan Paul, London, 1974.
7. *The Guardian,* 3 July 1967.
8. DEPARTMENT OF THE ENVIRONMENT, *Transport Policy: A Consultation Document,* HMSO, London, 1976, Vol. 1, p. 68.
9. For a comprehensive listing see F. W. S. CRAIG, *British General Election Manifestos 1900–1974,* Macmillan, London, 1975.
10. RICHARD CROSSMAN, *The Diaries of a Cabinet Minister:* Volume 1, *Minister of Housing 1964–66.* Hamish Hamilton and Jonathan Cape, London, 1975, p. 616.
11. TONY BENN, *The Guardian,* Agenda, 4 February 1980.
12. *The Guardian,* 30 October 1978 ("Civil servants back big lorries").
13. *The Times,* 7 October 1966 ("Mrs. Castle learns from U.S. mistakes").
14. PETER KELLNER and Lord CROWTHER-HUNT, *The Civil Servants: An Inquiry into Britain's Ruling Class,* Macdonald Futura, London, 1980.
15. J. J. RICHARDSON and A. G. JORDAN, *Governing Under Pressure: The Policy Process in a Post-Parliamentary Democracy,* Martin Robertson, Oxford, 1979.
16. WILLIAM PLOWDEN, *The Motor Car and Politics in Britan,* Penguin Books, Harmondsworth, 1973.
17. Quoted in JOHN WARDROPER, *Juggernaut,* Temple Smith, London, 1981.
18. RICHARD ROSE, *Do Parties Make a Difference?* Macmillan, London, 1980.
19. WILLIAM PLOWDEN, *The Motor Car and Politics in Britain,* Penguin Books, Harmondsworth, 1973.
20. M. E. BEESLEY, "Influence of Measures Designed to Restrict the Use of Certain Transport Modes", *Round Table* 42, European Conference of Ministers of Transport, Paris, 1979.

Illustrations

1.1. Adapted from J. M. W. STEWART, "A Pricing System for Roads", University of Glasgow Social and Economic Studies, *Occasional Papers* 4, Oliver & Boyd, Edinburgh, 1965, Table 2.
2.1. From information in JOHN TETLOW and ANTHONY GOSS, *Homes, Towns and Traffic,* 2nd ed., Faber & Faber, London, 1968, p. 83.
2.2. Birmingham City Council.
2.3. Keystone Press Agency.
2.4. From data in J. C. TANNER *et al.,* "Sample Survey of Roads and Traffic of Great Britain", *Road Research Technical Paper* 62, HMSO, London, 1962 (the distribution of traffic between road types in each region had been assumed to be the same as the nationwide distribution).
3.1 A. SAMUELS (Chairman), *Parking Survey of Inner London: Interim Report,* HMSO, London, 1956, Appendix 1.
3.2 Adapted from MINISTRY OF TRANSPORT, *Urban Traffic Engineering Techniques,* HMSO, London, 1965, Table 6.
3.3. Department of Transport.
3.4. Department of Transport.
3.5. OSBERT LANCASTER, *Graffiti,* John Murray, London, 1964.
3.6. *Transport Facts and Figures,* Greater London Council, 1973.
4.1. C. D. BUCHANAN *et al., Traffic in Towns,* HMSO, London, 1963.
4.2. C. D. BUCHANAN *et al., Traffic in Towns,* HMSO, London, 1963.
6.1. Adapted from *Bus Demonstration Project Summary Report 3,* Department of the Environment, 1973, Figure 1.

7.1. *Belfast Urban Motorway: Feasibility Report,* R. Travers Morgan and Partners, Belfast, 1967.
7.2. *London Traffic Survey,* Ministry of Transport and London County Council, 1964.
7.3. "Transport Now", *New Society* (1967) 640–1.
7.4. "Transport Now", *New Society* (1967) 640–1.
7.5. Adapted from *SELNEC Transportation Study: a Broad Plan for 1984,* Manchester, 1972, Figure 65.
8.1. Reading University collection.
8.2. *The Sunday Times,* 26 January 1969.
8.3. *Punch,* 2 August 1972.
9.1. R. A. VINCENT and R. E. LAYFIELD, "Nottingham Zones and the Collar Study – overall assessment", *Transport and Road Research Laboratory Report,* 805, Crowthorne, 1977, adapted from Figures 1 and 2.
10.1. National Bus Company, "Bus Priority Schemes", *Marketing and Operational Research Report* 19, Peterborough, 1978, Figure 7.
10.2. Adapted from data in National Bus Company, "Bus Priority Schemes", *Marketing and Operational Research Report* 19, Peterborough, 1978.
10.3. Adapted from *Transport Statistics Great Britain 1965–1975* (Table 20) and from *Transport Statistics Great Britain 1969–1979* (Table 59), HMSO, London.
10.4. Adapted from TREASURY, *The Government's Expenditure Plans 1979–80 to 1982–83* (Cmnd. 7439), HMSO, London, 1979, Table 2.6.
10.5. From information supplied by Leeds City Council.
10.6. Thames Valley Newspapers.
11.1. Adapted from Lord KINGS NORTON (Chairman), *Cars for Cities: A Study of Trends in the Design of Vehicles with Particular Reference to their Use in Towns,* Reports of the Steering and Working Group, HMSO, London, 1967, Figure 6.4.
11.2. *Punch,* 3 August 1977.
11.3. P. J. CORCORAN *et al.*, "Higher Gross Weight Goods Vehicles – Operating Costs and Road Damage Factors", *Supplementary Report* 590, Transport and Road Research Laboratory, Crowthorne, 1980, Figure 2.
11.4. Transport and Road Research Laboratory.
11.5. Leyland Vehicles Ltd.
12.1. From original data.
12.2. PHILLIP BLAKE, Local Government Operations Research Unit, Reading.
13.1. From data supplied by the Civic Trust.
13.2. Based on J. C. TANNER, "Forecasts of Future Numbers of Vehicles in Great Britain", *Roads and Road Construction* 40 (1962), 263–74 (Figure 12) and J. C. TANNER, "Forecasts of Vehicle Ownership in Great Britain", *Roads and Road Construction* **43,** (1965) 371–6 (Figure 9).
13.3. *Transport Statistics, Great Britain 1969–79,* HMSO, London.
13.4. Adapted from *Transport Statistics, Great Britain 1969–79,* HMSO, London.
14.1. From information supplied by the Department of Transport.

Appendix A
The Transport Portfolio 1945–1981

Ministers of Transport

3 August	1945	A. Barnes	(Labour)
31 October	1951	J. Maclay	(Conservative)
7 May	1952	A. Lennox-Boyd	(Conservative)

Ministers of Transport and Civil Aviation

1 October	1953	A. Lennox-Boyd	(Conservative)
28 July	1954	J. Boyd-Carpenter	(Conservative)
20 December	1955	H. Watkinson	(Conservative)

Ministers of Transport

14 October	1959	E. Marples	(Conservative)
16 October	1964	T. Fraser	(Labour)
23 December	1965	Mrs. B. Castle	(Labour)
5 April	1968	R. Marsh	(Labour)
5 October	1969	F. Mulley	(Labour)

Ministers for Transport (within the Department of the Environment)

15 October	1970	J. Peyton	(Conservative)
7 March	1974	F. Mulley	(Labour)
12 June	1975	J. Gilbert	(Labour)

Secretary of State for Transport

10 September	1976	W. Rodgers	(Labour)

Minister of Transport

5 May	1979	N. Fowler	(Conservative)

Secretary of State for Transport

8 January	1981	N. Fowler	(Conservative)

Index

Other Titles in the Series

CHADWICK, G. F.
A Systems View of Planning, 2nd Edition (Volume 1)

BLUNDEN, W. R.
The Land Use/Transport System (Volume 2)

GOODALL, B.
The Economics of Urban Areas (Volume 3)

LEE, C.
Models in Planning: An Introduction to the Use of Quantitative Models in Planning (Volume 4)

FALUDI, A. K. F.
A Reader in Planning Theory (Volume 5)

COWLING, T. M. & STEELEY, G. C.
Sub-regional Planning Studies: An Evaluation (Volume 6)

FALUDI, A. K. F.
Planning Theory (Volume 7)

SOLESBURY, W.
Policy in Urban Planning Structure Plans, Programmes and Local Plans (Volume 8)

MOSELEY, M. J.
Growth Centres in Spatial Planning (Volume 9)

LICHFIELD, N. *et al.*
Evaluation in the Planning Process (Volume 10)

SANT, M. E. C.
Industrial Movement and Regional Development: The British Case (Volume 11)

HART, D. A.
Strategic Planning in London: The Rise and Fall of the Primary Road Network (Volume 12)

STARKIE, D. N. M.
Transportation Planning, Policy and Analysis (Volume 13)

FRIEND, J. K. & JESSOP, W. N.
Local Government and Strategic Choice, 2nd Edition (Volume 14)

RAPOPORT, A.
Human Aspects of Urban Form (Volume 15)

DARIN-DRABKIN, H.
Land Policy and Urban Growth (Volume 16)

NEEDHAM, D. B.
How Cities Work: An Introduction (Volume 17)

DAVIDSON, J. & WIBBERLEY, G.
Planning and the Rural Environment (Volume 18)

FAGENCE, M.
Citizen Participation in Planning (Volume 19)

FALUDI, A. K. F.
Essays on Planning Theory and Education (Volume 20)

BLOWERS, A.
The Limits of Power: The Politics of Local Planning Policy (Volume 21)

McAUSLAN, J. P. W. B.
The Ideologies of Planning Law (Volume 22)

MASSAM, B. H.
Spatial Search: Applications to Planning Problems in the Public Sector (Volume 23)

HOYLE, B. S. & PINDER, D. A.
Cityport Industrialization and Regional Development (Volume 24)

TAYLOR, J. L. & WILLIAMS, D. G.
Urban Planning Practice in Developing Countries (Volume 25)

SPENCE, N. A. *et al.*
British Cities: An Analysis of Urban Change (Volume 26)

PARIS, C.
Critical Readings in Planning Theory (Volume 27)